John Howard Appleton

Lessons in chemical Philosophy

Second Edition

John Howard Appleton

Lessons in chemical Philosophy
Second Edition

ISBN/EAN: 9783337069339

Printed in Europe, USA, Canada, Australia, Japan

Cover: Foto ©ninafisch / pixelio.de

More available books at **www.hansebooks.com**

LESSONS

IN

CHEMICAL PHILOSOPHY.

BY

JOHN HOWARD APPLETON, A.M.,

Professor of Chemistry in Brown University,

AUTHOR OF

"Beginner's Handbook of Chemistry," "The Young Chemist,"
"Qualitative Chemical Analysis," "Quantitative Chemical Analysis."

Second Edition.

SILVER, BURDETT & CO., PUBLISHERS,
NEW YORK ... BOSTON ... CHICAGO.
1890.

PROFESSOR APPLETON'S WORKS ON CHEMISTRY.

I. THE BEGINNER'S HANDBOOK OF CHEMISTRY: *Price $1.00.* This is an introduction to the study of Chemistry, suitable for general readers. It treats chiefly the non-metals, these being generally found to furnish the best material for an elementary course, and to best illustrate the fundamental facts and principles of the science.

The book is written in attractive style, and has had a very large sale. It is profusely illustrated with engravings, and has, in addition, fourteen colored plates.

II. THE YOUNG CHEMIST: *Price 75 Cents.* A book of chemical experiments for beginners in Chemistry. This is designed for use in schools and colleges. It is composed almost entirely of experiments, those being chosen that may be performed with very simple apparatus. The book is arranged in a clear, systematic, and instructive manner.

III. QUALITATIVE ANALYSIS: *Price 75 Cents.* A brief but thorough manual for laboratory use.

It gives full explanations, and many chemical equations. The processes of analysis are clearly stated, and the whole subject is handled in a manner that has been highly commended by a multitude of successful teachers of this branch.

IV. QUANTITATIVE ANALYSIS: *Price $1.25.* A text-book for school and college laboratories.

The treatment of the subject is such that the pupil gains an acquaintance with the best methods of determining all the principal elements, as well as with the most important type-processes, both of gravimetric and volumetric analysis.

THE EXPLANATIONS ARE DIRECT AND CLEAR, so that the pupil is enabled to work intelligently *even without the constant guidance of the teacher.* By this means the book is adapted for self-instruction of teachers and others who require this kind of help to enable them to advance beyond their present attainments.

V. CHEMICAL PHILOSOPHY: *Price $1.40.* A text-book for schools and colleges.

It deals with certain general principles of chemical science, such as the constitution of matter; atoms, molecules, and masses; the three states of matter and radiant matter; the change of state from one form of matter to another. It also presents such topics as Boyle's and Mariotte's law, Charles's law, and the other general laws of matter. It discusses from a chemical standpoint certain forms of energy, such as heat, light, electricity. It treats of the nature of chemical affinity; the chemical work of micro-organisms; the modes of chemical action; thermo chemistry; and those attractions of substances which are partly physical and partly chemical. It also presents a full study of atomic weights, the methods leading to a first adoption of them, and then to the grounds sustaining certain numbers selected. The periodic system is of course discussed.

The work is fully illustrated.

Copies sent by mail, postpaid, by the Publishers, *upon receipt of the advertised price.*

COPYRIGHT, 1890,

BY JOHN HOWARD APPLETON.

TYPOGRAPHY BY J. S. CUSHING & CO., BOSTON.

PRESSWORK BY BERWICK & SMITH, BOSTON.

PREFACE.

This book is a formal presentation of certain subjects which the author has been in the habit of offering to his classes in the form of lectures. It is intended to explain to beginners, or even tolerably advanced students in chemistry, certain of the general laws of the science, and that in a compact and easily handled form. To a certain extent theories are given. While these have their important use, they must not be relied upon too strongly. "Theories," says Dumas, "are like crutches. To find out their value we must try to walk with them." On the other hand, where theories have been presented in this work, effort has been made to show distinctly the basis upon which they rest. Particularly in the chapters relating to atomic weight the attempt has been made to lead the pupil to formally distinguish between *facts* and *inferences*.

The efforts of investigators in chemistry as in other natural sciences are, as Berthelot remarks, to transform a mere descriptive science into a truly physical and mechanical, *i.e.* a mathematical, one. This thought has not been forgotten in arranging the work presented herewith. At the same time the attempt has been made to avoid making the book mathematical. The mathematical treatment, valuable as it is for highly advanced students, is apt to be repellent to beginners.

The author takes this opportunity to thank the teachers of the United States for the kind reception they have given to his earlier text-books in chemistry. He cherishes the hope that they may find this one of service to them in their studies and their teaching.

BROWN UNIVERSITY, PROVIDENCE, R.I.,
September, 1890.

CONTENTS.

                                                  PAGE

CHAPTER I. — The branches of natural science. The place of chemistry 1

CHAPTER II. — The constitution of matter. The atom. The molecule. (Elements and compounds.) The mass. The modern atomic theory 3

CHAPTER III. — Is matter indeed molecular and atomic? General discussion upon atoms and molecules. Ordinary observation. More searching examination. Evidences of heterogeneity found in mechanical and physical relations of substances; in their relations to heat, to light, and to the electric current; and in their chemical properties. Compound radicles. General conclusions. The atoms of the chemist viewed as composite. The genesis of atoms. Shapes of atoms. Movements of atoms. Positions of atoms in space 15

CHAPTER IV. — The three states of matter. Solids, liquids, and gases. Importance of the study of gases. Radiant matter . . . 35

CHAPTER V. — Change from one state of matter to another. Influence of addition and withdrawal of heat 42

CHAPTER VI. — Changes incident to addition of heat. Addition of heat to a solid. Rise of temperature according to specific heat. Melting. Latent heat of liquefaction. Special forms of liquefaction. Dissociation. Addition of heat to a liquid. Vaporization. Ice machines. Changes incident to withdrawal of heat. Withdrawal of heat from a gas. Withdrawal of heat from a liquid. Solidification of homogeneous and of mixed liquids . . . 45

CHAPTER VII. — Certain general laws of matter. Boyle's or Mariotte's law of the pressure of gases. Charles's law of the expansion of gases by heat. Graham's two laws of gaseous diffusion. The law of Henry and of Dalton of the relation of pressure to the solubility of a gas in water. The law of definite proportions. The two laws of multiple proportions. Gay-Lussac's three laws of combination of gases. Avogadro's and Ampère's hypothesis of the size of gaseous molecules 63

CHAPTER VIII. — Certain forms of energy closely connected with chemical change. Heat; temperature; expansion; change of state; chemical combination and decomposition; light (spectrum analysis); work. Electricity: its sources and effects . . 82

CHAPTER IX. — The attractions of masses. Gravitation . . . 99

CHAPTER X. — The attractions of molecules. I. Cohesion. In solids; in gases; in liquids. Polarity; crystallization; cleavage. Crystalline systems. The process of crystallization 101

CHAPTER XI. — The attractions of molecules (*continued*). II. Adhesion. (*A*) Adhesion between solids and solids. (*B*) Adhesion between solids and liquids; moistening; capillary attraction; spheroidal state; solution (deliquescence, freezing mixtures) . 114

CHAPTER XII. — The attractions of molecules (*continued*). II. Adhesion. The separation of a solid from a liquid. (*C*) Adhesion between solids and gases. (*D*) Adhesion between liquids and liquids. (*E*) Adhesion between liquids and gases. (*F*) Adhesion between gases and gases (the terrestrial atmosphere) . . 124

CHAPTER XIII. — The attraction of atoms. Chemical affinity. Conditions favoring chemical change: the liquid condition; heat (thermolysis and dissociation); light; electricity; vital processes of higher and lower living beings (organic and inorganic compounds) 141

CHAPTER XIV. — The attraction of atoms (*continued*). The chemical work of micro-organisms. Microbes: their conditions of growth; results of their action; their usefulness . . . 155

CHAPTER XV. — The attraction of atoms (*continued*). Modes of chemical action. Sphere of chemical action. Criteria of chemical action. Results of chemical action. General laws of chemical action, 168

CHAPTER XVI. — The attraction of atoms (*continued*). Thermochemistry: its laws and units. Calorimeters and the difficulties they have to meet. Range of thermo-chemistry. Results . . 175

CHAPTER XVII. — The attraction of atoms (*continued*). Theories of the nature of chemical attraction 187

CHAPTER XVIII. — Atomic weight: method of work and method of description 193

CONTENTS.

PAGE

CHAPTER XIX. — Atomic weight (*continued*). First step: A unit adopted. Second step: Selection of the compounds and the processes to be employed 199

CHAPTER XX. — Atomic weight (*continued*). Third step: Experimental work for securing a few atomic weights. Study of chlorine, bromine, and iodine; sodium, potassium, and silver . . 204

CHAPTER XXI. — Atomic weight (*continued*). Fourth step: The choice of a particular atomic weight from several combining numbers. Density of elementary gases. Volume composition of compound gases. Vapor density of compound substances. Specific heats of elements and of compounds. Atomic heats. Specific heats of chlorine, bromine, iodine, potassium, sodium, silver. A study of oxygen and of sulphur 209

CHAPTER XXII. — Atomic weight (*continued*). Fifth step: Confirmation of the atomic weights chosen. Molecular formula supported by volume composition, by chemical substitution, by melting-points and boiling-points, by crystalline form, by molecular stability, by relationship, by results of decomposition, by exceptional compounds, by special properties of substances . . . 225

CHAPTER XXIII. — Atomic weight (*continued*). Sixth step: Bring all the atomic weights into one table. (The periodic law.) The work of Newlands, Mendeléeff, and Carnelley. Prout's hypothesis, 243

CHAPTER XXIV. — Atomic weight (*continued*). Elementary substances as molecular 249

CHAPTER I.

THE BRANCHES OF NATURAL SCIENCE.

THE PLACE OF CHEMISTRY.

THERE may be as many sciences as there are kinds of subject-matter for scientific treatment.

The scientific treatment of a subject demands exact observation, precise description with fixed nomenclature, classified arrangement, rational explanation.

The term *natural science* is usually applied to the classified knowledge of external material nature and certain of its forces.

In the ordinary every-day use of language, then, the general term *science* is often applied to what is here included in the term *natural science*. But it must not be forgotten that there may be a science of the human mind, for example, as well as sciences of external forms of matter.

Divisions of Natural Science. — One grand division of natural science is that called *Natural History*. In this are included —

Geology, a history of the inanimate matter of the earth, and embracing physical geography and meteorology;

Zoölogy, a history of animals;

Botany, a history of vegetable beings.

Natural history is mainly descriptive.

Another grand division of natural science is that called

Natural Philosophy, or *Physical Science*. In this are included —

Mechanics, which treats of masses of matter;

Physics proper, which treats of the motions of molecules, and the molecular forces such as light, heat, and electricity;

Chemistry, which treats of atoms, the constitution and properties of molecules, and the laws of chemical change.

Physical science is mainly explanatory.

Defects of the Foregoing Classification. — Even a very hasty consideration of this brief classification shows that, especially as respects exact lines of demarcation, it is inadequate. Thus the individuals discussed under each department of natural history involve in their histories the processes of natural philosophy; for the animal, the plant, and the rock are formed, or live, or grow, or merely exist in one place, as the case may be, under conditions involving mechanical, physical, and chemical forces.

Again, it will be noted that certain branches of study evidently belonging to natural science — astronomy, for example — are not specifically mentioned.

But the defects of classifications of this sort are referable to difficulties that nature itself places in our way, for the multitude of natural phenomena are not in themselves characterized by strongly marked division lines, but are mostly intimately interwoven.

Finally, it is not intended, in this chapter, to offer a perfect classification of the subjects of study afforded by natural objects and forces; it is merely proposed to place before the reader the general relations of chemistry to other natural sciences.

CHAPTER II.

THE CONSTITUTION OF MATTER.

THE ATOM; THE MOLECULE; THE MASS.

MATTER is believed to be capable of existing in portions of three different grades of magnitude. These are called respectively the *atom*, the *molecule*, the *mass*.

The Atom. — An atom is the smallest unit of matter *now recognized* as existing. About seventy different kinds of atoms are now known.

Each atom of matter is viewed as possessing the following characteristics, in addition to many others : —

It is extremely small (but not *infinitely* small).
It is indivisible, and, indeed, in itself unchangeable.
It possesses a definite weight, which may be determined relatively and absolutely.
(The atomic weight is different for different kinds of atoms, but practically the same for atoms of the same kind. Thus each atom of hydrogen weighs 1 *microcrith*; each atom of oxygen weighs about 16 *microcriths*. See p. 199.)
It is capable of manifesting an attractive force, called *chemical affinity*.
It almost invariably exists in a group, of which the component atoms may be alike or may be unlike. In a very few cases an atom may exist singly.

Examples of single atoms are : —

    Barium (Ba),
    Cadmium (Cd),
    Mercury (Hg) (*Hydrargyrum*),
    Zinc (Zn).

The Molecule. — A molecule is the smallest particle of any substance that manifests the chemical properties of that particular substance. Thus: —

H_4C represents one molecule of marsh gas.
H_3N " " " " ammonia gas.
H_2O " " " " water.
H_2 " " " " hydrogen.
O_3 " " " " ozone.
O_2 " " " " ordinary oxygen.

A molecule is believed to be capable of possessing most of the following characteristics, in addition to many others: —

(a) It is extremely small. From recent physical investigations, "it may be concluded with a high degree of probability that in ordinary liquids or solids the diameter of the molecule," that is, the distance between the centres of contiguous molecules, is between the one two-hundred-and-fifty-millionth $\left(\frac{1}{250,000,000}\right)$ and the one five-thousand-millionth $\left(\frac{1}{5,000,000,000}\right)$ of an inch.

(b) It is not completely nor absolutely in contact with its neighboring molecules, but is separated from them by relatively large spaces.

(c) When in the state of gas a molecule demands the same amount of space as every other molecule in the gaseous state (under the same conditions of temperature and pressure).

(d) A molecule usually consists of a group of atoms. These atoms are bound together by chemical affinity and into a union of exceedingly intimate relationship. In fact, in most cases the molecule cannot be divided without absolutely changing the identity of the substance. This result might be expected in the case of molecules composed of different kinds of atoms.

Thus water has molecules, each expressible by the formula H_2O. When water is decomposed, the change takes place as follows: —

$2\,H_2O$,	*when decomposed, yield*	$2\,H_2$	$+$	O_2
Two molecules of Water,		Two molecules of Hydrogen gas.		One molecule of Oxygen gas,
36 parts by weight.		4 parts by weight.		32 parts by weight.
36			36	

But though at first unexpected, it is easily seen that a change of identity may result from the decomposition or rearrangement even of molecules whose individual atoms are of precisely the same kind. Thus ozone has molecules, each expressible by the formula O_3. When ozone is decomposed, the change takes place as follows: —

$2\,O_3$	*yield by decomposition*	$3\,O_2$
Two molecules of Ozone,		Three molecules of Ordinary Oxygen,
96 parts by weight.		96 parts by weight.

Now it is an observed fact that the properties and powers of ordinary oxygen are very different from those of ozone.

(*e*) The term *molecule*, usually reserved for groups of atoms, is extended to single atoms in the four exceptional cases, already cited, in which the single atom is capable of existing apart. Thus a molecule of mercury means also a single atom of mercury.

Elements and Compounds. — Matter is called *simple*, or *elementary*, when its molecules are composed of atoms of the same kind.

Most of the elementary gases are believed to have two atoms in the molecule.

Thus, the hydrogen molecule is expressed

$$H - H \text{ or } H_2,$$

and the oxygen molecule,

$$O - O \text{ or } O_2.$$

Matter is called *compound* when its molecules are made up of different kinds of atoms. The number of atoms in many compound molecules is but two; in others it is greater, sometimes reaching to several hundreds.

Examples: Starch, $C_6H_{10}O_5$,
                           21 *atoms.*
               Protagon, $C_{116}H_{201}N_4PO_{22}$,
                           434 *atoms.*

The Mass. — A *mass* of matter is a collection of molecules. Portions of matter appreciable by the senses are, in most cases, masses.

FIG. 1.— Antoine Laurent Lavoisier. Born in Paris, August 26, 1743; died on the scaffold in Paris, May 8, 1794.

The Modern Atomic Theory. — The views of the constitution of matter set forth in the foregoing paragraphs may be said to have their origin in the atomic theory of Dr. John Dalton, an English mathematician, who began to develop his theory in the year 1803.

In some ancient metaphysical speculations matter

was held to be infinitely divisible, while in others the contrary view was maintained.

The ancient philosophers who constructed atomic theories felt sure, upon general grounds, that matter is susceptible of division to a degree far beyond that which

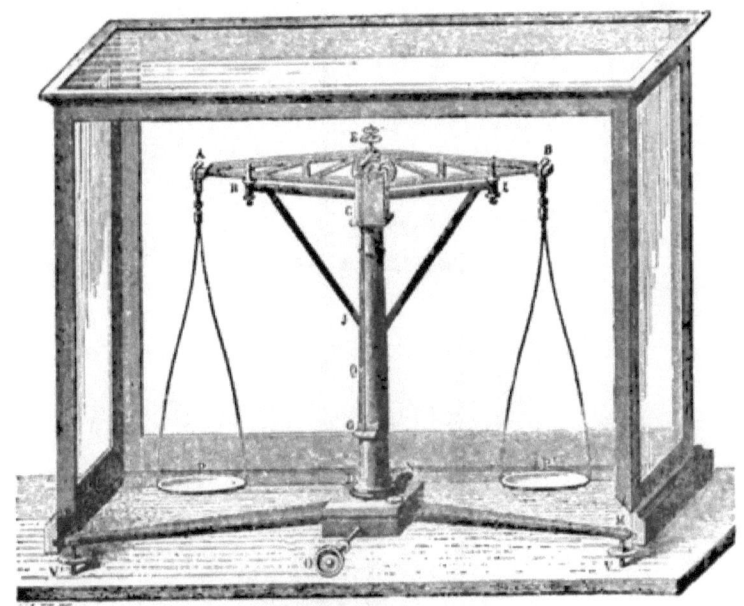

FIG. 2.— Chemical balance. (A portion is shown cut away so as to display the tripod.)

their appliances effected. But when the question arose, "Is not matter then infinitely divisible?" no valid answer could be given. An unconquerable difference of opinion existed.

The ancient views were defective because they were too largely speculative and were not based on a sufficiently large number of facts — especially such facts as can be learned only by carefully devised and conducted experiments.

The modern chemist declares that in fact there exist at present limits to the divisibility of matter. Moreover, Dalton's theory, as well as other modern views of the constitution of matter, are based not upon specula-

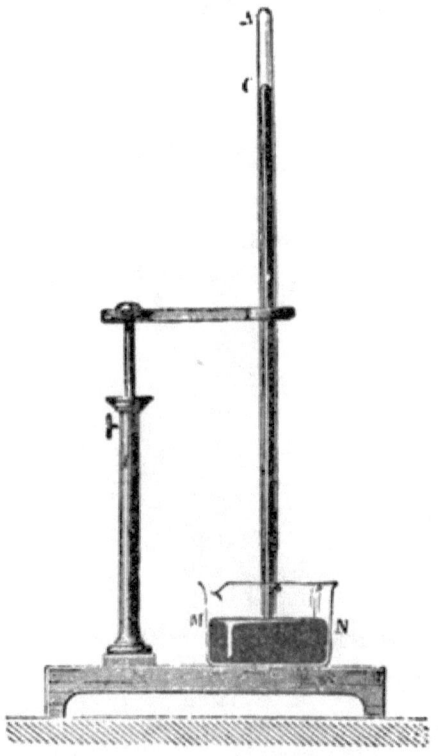

FIG. 3.— Simple form of barometer, called Torricelli's. The tube A is closed at the top. The mercury does not fall because of the atmospheric pressure exerted upon the surface of mercury at MN.

tion, but upon discoveries reached by accurate experiments with carefully devised appliances.

The most important and suggestive facts were collected by the chemists of the last part of the eighteenth century, and especially Lavoisier. He led the way by

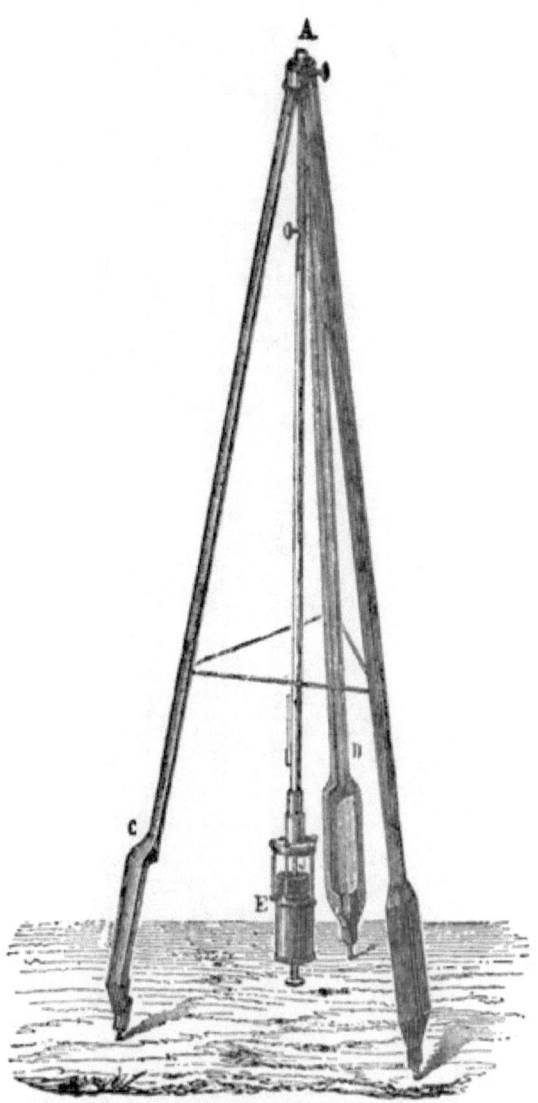

FIG. 4. — Special barometer used for measuring the heights of mountains. The tripod is constructed so that it may be closed, and thus protect the more delicate parts of the barometer. Upon ascending a mountain the diminished atmospheric pressure is manifested by the fall of mercury in the long tube of the barometer.

Fig. 5. — Water barometer in the Tour Saint-Jacques, Paris. Of course the tube is much longer than a mercurial barometer. On the other hand, the variations in the height of the barometer are much more marked. A variation of atmospheric pressure representing an inch on the mercurial barometer produces a variation of about 13.6 inches in the water barometer.

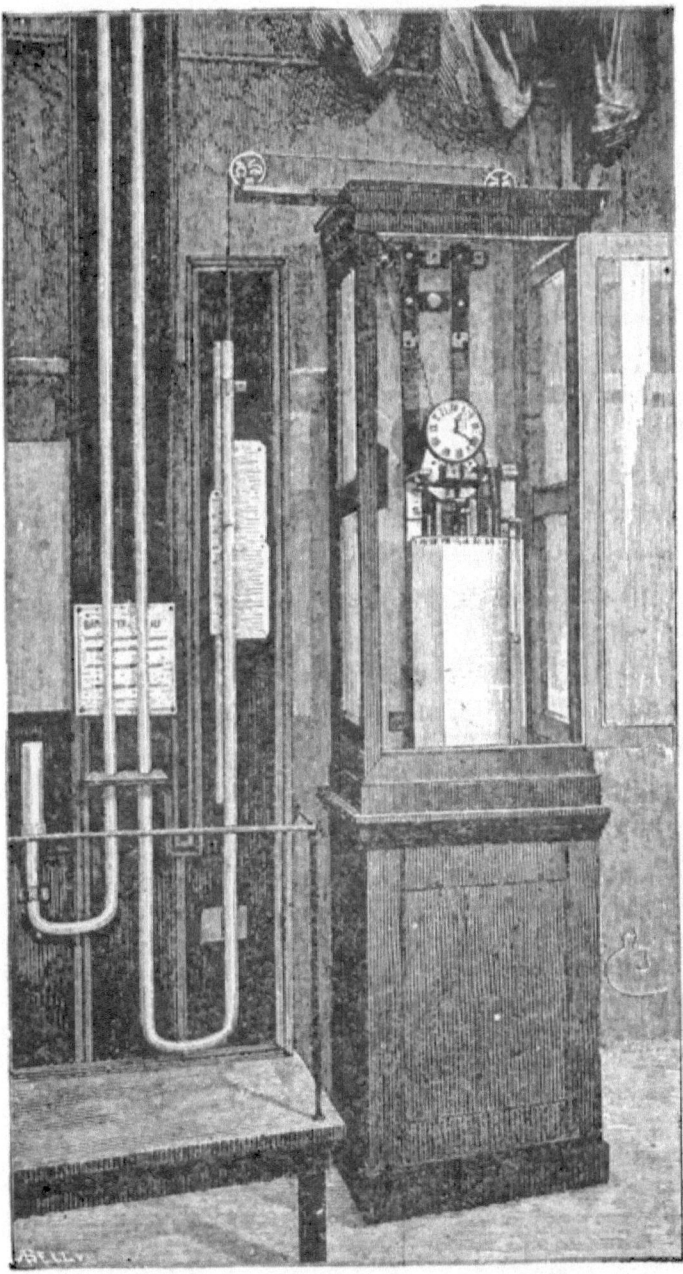

Fig. 6. — Larger view of the base of the water barometer shown in section in Figure 5. The self-registering apparatus containing clock-work and paper cylinder is shown at the right.

his rigid use of the balance in chemical investigation. He maintained — what was not perceived before — that *weight* is an important function of matter, and a safe guide in chemical reasoning. His teaching showed the value of those quantitative methods which have not only afforded a sure basis for modern theories of matter, but have been a most important aid in the general modern progress of chemistry.

The first notions of the modern atomic theory appear to have been suggested to Dalton by following Lavoisier's methods; *i.e.* by the use of the quantitative processes of the time. Dalton's work led also to the enunciation of the important laws of *definite proportions* and of *multiple proportions*, — laws which are still the chief support of that theory.

Modern instruments of precision, notably the balance, the barometer, the graduated eudiometer,[1] and the thermometer have afforded large contributions toward new and exact knowledge of matter and its forces; they have also aided to dispel many old and false impressions.

"The vicissitudes in the fortunes of a truly scientific idea are aptly illustrated by the history of the atomic theory. After a period of dormancy of more than 2000 years the atomic theory was revived and rendered definite by Dalton, was firmly established on an experimental basis by Berzelius, was almost abandoned by the school founded by the same chemist, was rehabilitated and again nearly despaired of by Dumas, was largely advanced by Avogadro, was subdivided and its parts clearly distinguished by Gerhardt and Laurent, and is now the foundation-stone of a great and ever-increasing edifice."[2]

Atoms and molecules are now considered to be real existences as truly as are planets and fixed stars; and they are as truly susceptible of measure-

---

[1] Eudiometer: an instrument or vessel for exact measurement of the volume of a gas.

[2] Muir, Principles of Chemistry, p. 24.

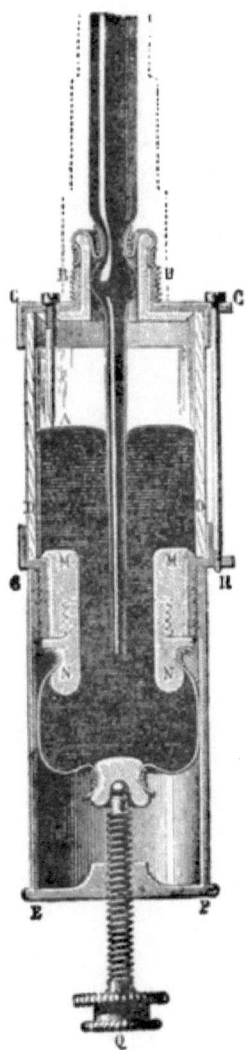
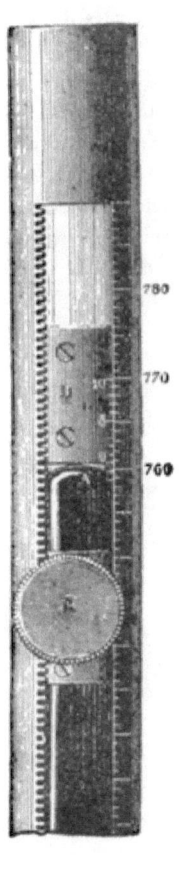

FIG. 7. — Portion of a barometer, showing at the top a part of the long glass tube containing mercury. The reservoir of mercury in the tube may be protected from undue agitation by advancing the screw Q so that its plug may close the lower aperture of the long glass tube.

FIG. 8. — Top of mercury column of a barometer. The rack and pinion movement is for the purpose of moving the vernier in order to afford more accurate reading.

ment. Their motions, though of a different kind, are as much matters of fact.

It must be admitted that even to-day some persons question the existence of the chemical atoms. The difficulty is probably the outcome of ambiguity in the word *atom*. The atom of the Greek atomists may or may not be the atom of the modern chemist. The Greek atom is something that by its very nature cannot ever be divided. The atom described by modern scientists is merely a very minute portion of matter that has not yet been divided. The Greek atom is a metaphysical creation; the atom of modern science is a real thing.

CHAPTER III.

IS MATTER INDEED MOLECULAR AND ATOMIC?

EVIDENCE POINTING TO CERTAIN DEGREES OF HETEROGENEITY.

General Discussion upon Atoms and Molecules. — Probably many persons — and especially those whose reasoning powers are but imperfectly developed — would have more confidence in the existence of atoms and molecules if they were visible.

But after all, the senses are by no means infallible guides. The sense of sight often deceives; even in the commonest affairs of life reliance is often placed upon the reasoning powers in opposition to the direct evidence of the eye. Especially when the eye is diseased, it gives false ideas of color, form, and the like. Even excellent eyes may not be able to see the vibration of a certain piano-string while it is singing; but by application of processes of reasoning to experimental results proof can be obtained, to the satisfaction of every one, of the existence of such vibration. It is only by processes of reasoning that the distances of the heavenly bodies are measured, but no one doubts the data that astronomers furnish.

However, single atoms and single molecules are not visible, and, moreover, it does not seem likely that optical appliances will ever extend the range of man's visual powers to such minute objects. But neither these difficulties nor any others need deter the investigator from efforts to solve the problem of the constitution of matter. He may with propriety continue to ask such general questions as the following : Is matter constructed on the type of an imperforate mass, or is it more truly a network of some sort? Does one minute portion of a given kind of substance come into absolute contact with its neighboring portions, so that no pointed implement, however fine and sharp, can discover places of more and of less resistance, or are there minute avenues of some sort affording more easy passage in some places than in others? Is matter indeed atomic and molecular, as has been declared?

The questions come to this : Is matter heterogeneous? If the answer is yes, the next questions are as to the character and extent of this heterogeneity.

These questions do indeed receive decisive answers from physical and chemical facts. By proper reasoning from natural phenomena there have been secured, at first approximate, and later highly exact, answers. A blind man may discover the construction of a gate that bars his path. By applying to such an obstacle instruments of various shapes and sizes he may be able to state whether it has the apparent continuity of sheet-iron, or has horizontal or vertical bars, or is of netted gauze, or even has perforations like fine card-board. In like manner scientific observations may be made as to the action of matter in response to certain general experiments. Then proceeding on and on, there may be secured a tolerably certain knowledge of its constitution.

Ordinary Observation upon this Subject. — Some forms of matter show plainly to the eye that they are made up of more than one kind of substance — that they are heterogeneous. Other kinds may appear to the eye to be homogeneous, while a fuller knowledge of them makes it

IS MATTER INDEED MOLECULAR AND ATOMIC? 17

absolutely certain that they are not so. As a very simple example consider sugar. It appears at first consideration to be alike all through; it is very easy to prove, however, that it contains at least carbon — something very different from sugar.

More Searching Examination. — There is presented below a very brief résumé of certain observed properties or actions of matter that point first to general heterogeneity or grained structure of some sort, and next to distinct molecular, and finally to distinct atomic composition.

The facts stated here are by no means all that apply. The whole body of physical and chemical knowledge points in this direction.

FIRST. *Evidences of heterogeneity found in certain mechanical properties of substances.*

(*a*) The diminution in volume of solids, of liquids, and — in a yet more marked degree — of gases, by *mechanical pressure*, leads to the suggestion that there are interior spaces which are diminished by that action.

(*b*) In similar manner peculiarities of internal structure are revealed by certain changes of the outline of bodies. The various degrees of *elasticity, hardness, malleability,* and *ductility* displayed by various solids show various kinds of heterogeneity.

<small>The changes of thickness of soap-bubble films, before they break, have been carefully studied. The relations of these films to light show their thickness at different times. The thinnest films appear to be commensurable with the diameters of a molecule, as learned by other methods.</small>

(c) Crystals possess *cleavage* and certain other properties which offer most marked suggestions of internal peculiarities of structure.

The foregoing classes of facts point, therefore, to general heterogeneity of substances in their internal make-up.

SECOND. *Evidences of heterogeneity found in certain physical properties of substances.*

(a) *Gases* diffuse readily into other gases. Liquids readily absorb gases. Certain solids swallow large quantities of gases without corresponding increase of bulk. (Palladium and platinum occlude hydrogen in an especially noteworthy degree.)

(b) *Liquids* diffuse themselves into other liquids with great rapidity, sometimes (as in case of alcohol and water) with diminution of volume.

(c) *Solids* dissolve quickly in liquids without unusual external influence, diffusing themselves throughout the solvents. (Dialysis is a specialized form of the motion of solids into liquids.)

Some solids seem to dissolve in gases, rising up into those gases under certain conditions, at temperatures far below those at which these solids ordinarily volatilize.[1]

These classes of facts point to internal places of feeble resistance in substances. Thus they point to molecular grouping of the firmer portions.

THIRD. *Evidences of heterogeneity found in the relations of certain substances to heat.*

---

[1] Hannay and Hogarth.

IS MATTER INDEED MOLECULAR AND ATOMIC? 19

(*a*) Solids, liquids, and gases expand upon addition of heat and contract upon its withdrawal. The details of these operations need not be presented here. But the fact that different gases expand equally with heat is a

FIG. 9. — The flask contains a dark-colored liquid, which does not allow the light to pass. It is diathermanous, however, for heat rays go through it easily, as may be demonstrated by concentrating them upon a piece of phosphorus or other combustible material.

special case under this head, and it points to *similarity* of arrangement of particles in gases.

The fact that some crystals expand more in certain directions than in others adds force to the argument; so do the facts connected with changes of substances in what may be called an upward direction from solids to liquids and gases, and downward from gases to liquids and solids, respectively by addition or withdrawal of heat.

The continuity of the three states of matter as shown by Thomas Andrews furnishes additional general evidence of molecular structure in matter. (See p. 35.)

(*b*) When certain substances are highly heated, they acquire anomalous vapor densities. These point to dis-

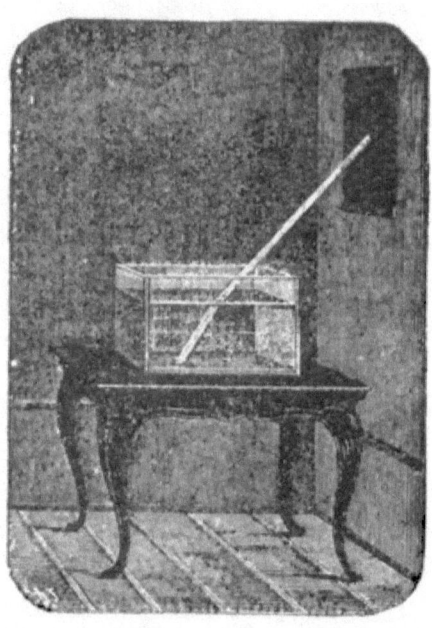

FIG. 10.—Arrangement for showing the refraction of light by water. A beam of light, coming through the aperture, penetrates the water in the jar, but is bent out of its original course.

sociation of molecules (and thence to the existence of molecules).

(*c*) All substances when highly heated give out light. In most cases a given metal, for example, gives out light made up of waves having many different rates of vibration. This fact is proved by the spectroscope.

Moreover, the same substance may give different colors (and so different spectra) at different high temperatures.

These facts seem to prove that an apparently homogeneous substance may possess *different internal vibrating parts.* (See p. 86.)

(*d*) Certain bodies manifest in a remarkable degree the phenomena associated with specific heat and latent heat. (See pp. 45 and 47.) Again, as respects radiant

FIG. 11. — Crystal of Iceland spar, showing double refraction; the single line is made to appear double.

dark heat, some substances are distinctly diathermanous, while others are athermanous. These peculiarities cannot be discussed at length here. But they are explicable only upon the theory that the substances displaying them are molecular. Thus they support the proposition under review.

FOURTH. *Evidences of heterogeneity found in the relation of certain substances to light.*

22 IS MATTER INDEED MOLECULAR AND ATOMIC?

The optical properties of substances afford some of the most clear and decisive evidences of *grainedness* of structure in bodies. The rays of light appear to be capable of use as a penetrating agency of a most discriminating character.

It is not practicable here, however, to do more than allude to the classes of optical facts referred to.

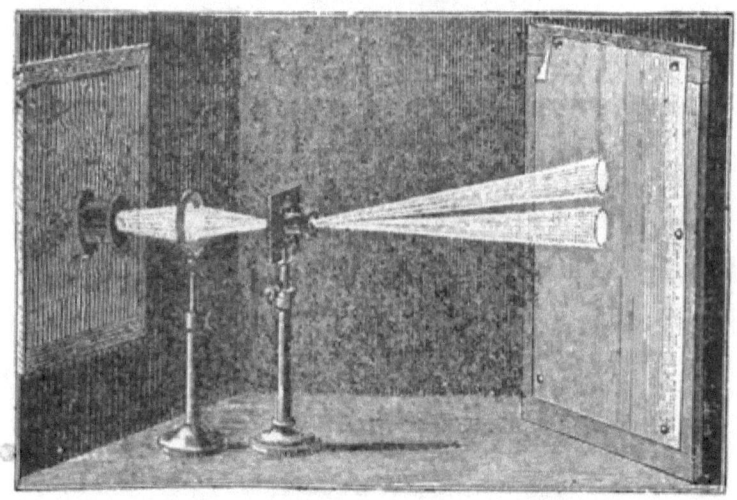

FIG. 12.— Method of displaying phenomena of double refraction. A single beam of light passed through a double-refracting crystal appears on the screen as two beams of light.

The opacity, the translucency, or the transparency of a given body; the phenomena of ordinary and of double refraction; the polarization of light, and the passage of the polarized ray through crystals in certain positions and not in others; the peculiar passage of the polarized ray also through solids under pressure as compared with the same at normal conditions; the rotation of the polarized ray by sugar solutions and the like, — all contribute greatly to substantiate the molecular theory. (In case of certain substances, as cane sugar, the greater the quantity of the substance, the greater the rotation. This suggests some greater interference of a greater number of molecular particles.)

IS MATTER INDEED MOLECULAR AND ATOMIC? 23

FIFTH. *Evidences of heterogeneity found in the relations of certain substances to the electric current.*

(*a*) In some cases elongation takes place when substances are rendered magnetic by an electric current. Thus iron bars and steel bars are so elongated.

(*b*) A certain substance (boro-silicate of lead) transmits polarized light when electrified, otherwise not.

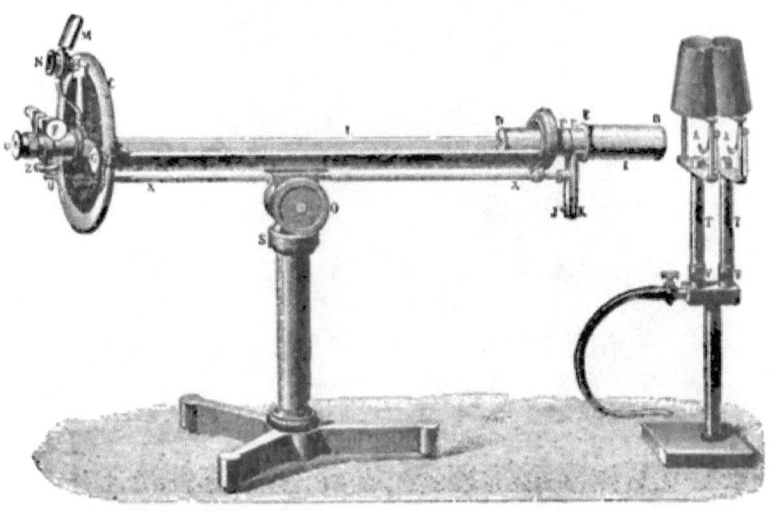

FIG. 13. — Polarizing saccharimeter. A ray of polarized light passing through a solution of cane sugar in water is rotated in such a way as to enable the observer to determine the percentage of sugar in the sample.

(*c*) Certain rarified gases and vapors show peculiar stratification when made luminous by the electric discharge of the Ruhmkorff coil. The stratification is different for different gases. This indicates the existence of internal *portions* in these substances.

(*d*) A given portion of oxygen gas is changed by the silent electric discharge into ozone without loss of weight, but with diminution of volume.

(e) The phenomena of electrolysis are very striking. A given battery current passing simultaneously through several compounds in different vessels may decompose them all at once. The weights of certain elements thus separated are very different. The following examples may be given: —

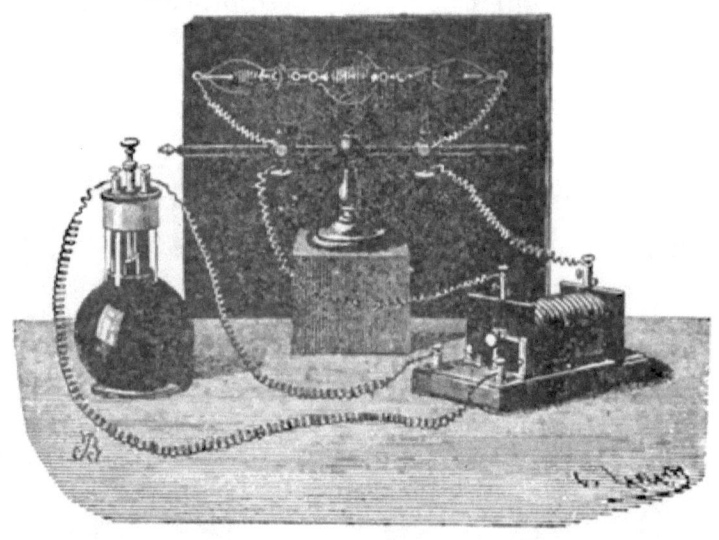

FIG. 14. — A portion of gas appears at first homogeneous, but is shown in disconnected strata when under the influence of an electric spark furnished by a Ruhmkorff coil. The coil is connected with a single cell of a bichromate battery. The glass tube containing the gas is called a Geissler tube.

Hydrogen	2.0	parts by weight.
Copper	63.2	" " "
Zinc	64.9	" " "

The amounts are closely proportioned to the amounts separated by similar *chemical* operations. These and other similar numbers represent a tendency to subdivision by different but definite quantities. Thus they show definite internal *structure* in masses of the substances mentioned.

General Comment on the Foregoing Evidences. — Heat, and especially light and electricity, are agencies possessing an extraordinary power of penetrating bodies. So far as they penetrate them with greater or less ease in certain directions or under special conditions, they cannot fail to raise the suggestion that the bodies in question are not perfectly homogeneous, but possess spaces of greater or less resistance. But this is what was to be demonstrated by the foregoing paragraphs.

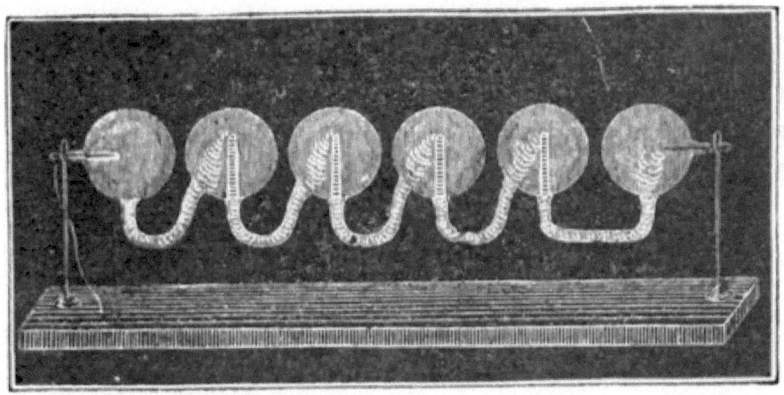

FIG. 15. — Geissler tube, showing that a gas appearing at first to be homogeneous manifests stratification when illuminated by an electric current from a Ruhmkorff coil.

SIXTH. *Evidences of heterogeneity found in the chemical relations of substances.*

The *chemical* suggestions of molecular structure are far more distinct and precise than the mechanical and physical. These may now be discussed. Of course only a few will be presented. If completeness were sought, the entire body of chemical knowledge would have to be adduced. The whole of modern chemistry must be referred to as the argument in this part of the case.

Illustration. — Chemical analysis shows that most substances are made up of more than one kind of matter. With the exception of about seventy substances called elements, everything known has internal complexity instead of internal uniformity throughout.

Thus the purest cooking salt, apparently homogeneous, is really made up of two very different substances or kinds of matter, — chlorine and sodium.

The chlorine has certain properties — distinctly its own — of which may be mentioned its gaseous form, its green color, and its bleaching power. And similarly, the sodium has certain properties — distinctly its own — of which may be mentioned its metallic lustre, its affinity for oxygen, and its power to decompose water in a certain way.

By *analysis* pure salt may be subdivided into portions of these two very different kinds of matter.

And by *synthesis* chlorine and sodium may be made to combine, and when they do so combine, they form a new product, which is precisely common salt and nothing else.

But the new product is not a *mere mixture* of its constituents.

(*a*) The constituents are more thoroughly intermingled or interdiffused than is possible in mere mixtures. For the minutest portions that can be recognized as containing true salt contain both of the factors, and not a single one.

(*b*) The constituents are associated in fixed and definite proportions by weight (always 23 parts of sodium to 35.4 parts of chlorine), — facts of which mere mixtures are independent.

(*c*) The constituents cannot be separated by any mechanical processes nor by any processes other than chemical or physico-chemical.

(*d*) The new product, the common salt, has no one of the six properties heretofore enumerated as possessed by the constituents (nor yet has it the *average* of these properties): instead, it has a new set of properties entirely its own.

It is plain that a mass of common salt is made up of small portions of salt, each small portion containing a definite amount of chlorine and of sodium. These small portions are the units already called molecules. Each of these molecules is made up of heterogeneous parts. But this is the point that was to be demonstrated.

Practically the same method of reasoning may be applied to every other substance known, except the seventy elementary ones. But further, a careful study of the elementary substances has shown that these are composed of the little groups called molecules, only in their cases the heterogeneity of parts is far less marked in quantity and quality, though apparently equally marked in a *reality* of distinct portions.

Compound Radicles. — Sometimes a group of atoms is recognized as capable of transfer as a whole from one complex molecule to another, and yet not capable of existing alone. To such a group the name compound radicle is applied. A simple example is found in the hypothetical metal ammonium (NH_4). Thus the compound called sal-ammoniac (ammonic chloride, NH_4Cl), contains the group NH_4, which is capable of continued transfer, as a whole, from one molecule to another; it acts very much like an element — only, if liberated, it at once decomposes. Chemists know a large number of other compound radicles of similar type; indeed, a whole class of bodies called compound ammonias has been studied very thoroughly by Hofmann and others.

Conclusions Reached. — There can be no reasonable doubt then that there exist molecules — in the general sense of *small portions of matter that preserve an integrity through a series of operations.* They are in fact units of a certain order.

Further, there can be no doubt that the molecular

FIG. 16. — August Wilhelm Hofmann, Professor of Chemistry in the University of Berlin.

units are made up of units of another order, — the so-called elementary atoms. These units are also integral parts, and they preserve their unbroken composition — whatever it may be — in spite of any processes as yet applied to them. Whether or not they are real atoms, in the sense of absolutely indivisible things, is not positively known. *There are certain grounds for believing that they are not.* That they are units of a most important character must be true. Their recognition is the work of modern investigators. Their discovery is a most valuable and fruitful one, whatever theory of their ultimate composition any one may hold, or whatever new facts relating to them may be discovered hereafter.

Are the Atoms of the Chemist Composite? — The notion has often been presented that the seventy substances called elementary are in fact very stable compounds not yet decomposed. Chemists, physicists, and philosophers have explicitly declared their belief in this view. Dalton, Faraday, Gladstone, Brodie, Graham, Mills, Stokes, Lockyer, and other experimenters have distinctly held it. From another side, thinkers like Herbert Spencer have recognized the necessity of it.

Four general grounds for believing the elements to be composite may be stated here: —

FIRST. Lockyer considers that the fact that a given element affords a multitude of spectrum lines at once suggests that these lines are somehow connected with different vibratory powers of different portions of the atom. In other words, the atom is composite.

SECOND. Dr. Carnelley offers a strong argument from the analogies of certain acknowledged compounds. It is this: There are recognized, in the group of hydrocarbons, a continuous series of compounds advancing in molecular weights by regular numerical steps, and having, with these, a regularly advancing sequence of physical and chemical properties. So the elements, as arranged by Newlands, Mendeléeff, Lothar Meyer, and others, constitute a series of substances whose members show a regular advance in atomic weights, and with it a regular progress of physical and chemical properties. There is no doubt that the hydrocarbons are compounds; there is then a high probability that the elements are so also. Carnelley goes so far as to offer the theory that the seventy elements are composed by union, in different ways, of *three kinds of ultimates.*

THIRD. Professor Crookes considers that he has in certain cases actually decomposed certain elements. He says that the substances he has obtained from so-called "old yttria" "are not *impurities* in yttrium. They constitute a veritable splitting up of the yttrium molecule into its constituents."

FOURTH. Certain so-called elements, as tellurium and didymium, have been considered compounds on grounds based on their general chemical properties.[1]

The Genesis of Atoms. — Professor Crookes proposes the theory that matter, as we now know it, has been produced by a kind of evolution from an antecedent substance or fire-mist which he calls *protyle*. He considers that by a series of large falls of temperature alternating with periods of approximately stationary temperature, the *protyle* has become condensed or combined into a series of substances — our elements.

"The first-born element would, in its simplicity, be most nearly allied to *protyle*. This is hydrogen, of all known bodies the simplest in structure and of the lowest atomic weight.

"Any well-defined element may be likened to a platform of stability, connected by ladders of unstable bodies. In the first coalescence of the primitive stuff there would be formed the smallest atoms; these would then unite, forming larger groups; the gaps between the several stages would gradually be bridged over, and the stable element appropriate to that stage would absorb, so to speak, the unstable rungs of the ladder which led up to it.

"In this genesis of the elements the longer the time taken up in the cooling-down process, during which the hardening of *protyle* into atoms takes place, the more sharply defined would be the resulting elements; whilst the more rapid and the more irregular the cooling, the more closely the resulting bodies would fade into each other by almost imperceptible degrees. Thus we may conceive that the succession of events which gave rise to such groups as platinum, osmium, and iridium — palladium, ruthenium, and rhodium — iron, nickel, and cobalt — might have produced only one element in each of these three groups if the process had been greatly

---

[1] Brauner, J. Ch. Soc., 1889, p. 382.

prolonged. And conversely, had the rate of cooling been much more rapid, elements might have originated still more nearly identical than are nickel and cobalt. Thus may have arisen the closely allied elements of the cerium, yttrium, and similar groups. In fact, we may regard the collocation of the

FIG. 17. — Sir William Thomson, distinguished English electrician and physicist.

minerals of the class of samarskite and gadolinite as a kind of cosmical lumber-room, where elements in a state of arrested development — the unconnected missing links of inorganic Darwinism — are gathered together."

Shapes of Atoms. — Sir William Thomson has propounded the theory that atoms possess the shape of rings of various kinds. In this view each ring possesses a vortex motion, something like that produced by the spontaneous ignition of bubbles of phosphuretted hydrogen.

Movements of Atoms. — It is at present unknown whether there is any constant motion of atoms within the molecule unattended by decomposition of it. General analogy and certain facts strongly favor the affirmative view. There is plenty of room in gaseous molecules at least.

Of course when a molecule is decomposed there is generally necessity of movement of its atoms; but this kind of motion may be only temporary, dependent upon the duration of the particular exciting cause.

Relative Position of Atoms in Space. — The brilliant studies of Kekulé upon organic compounds have given rise to pretty general adoption of a certain expression for the substance benzol (C_6H_6). This takes the form of a hexagon called the benzol hexagon or the benzol ring, as represented by the diagram below: —

$$\begin{array}{c}
\mid \\
C \\
\diagup \diagdown \\
-C \qquad C- \\
\mid \qquad \parallel \\
-C \qquad C- \\
\diagdown \diagup \\
C \\
\mid
\end{array}$$

In this expression the six atoms of carbon are viewed as forming a closed chain, of which each atom is attached

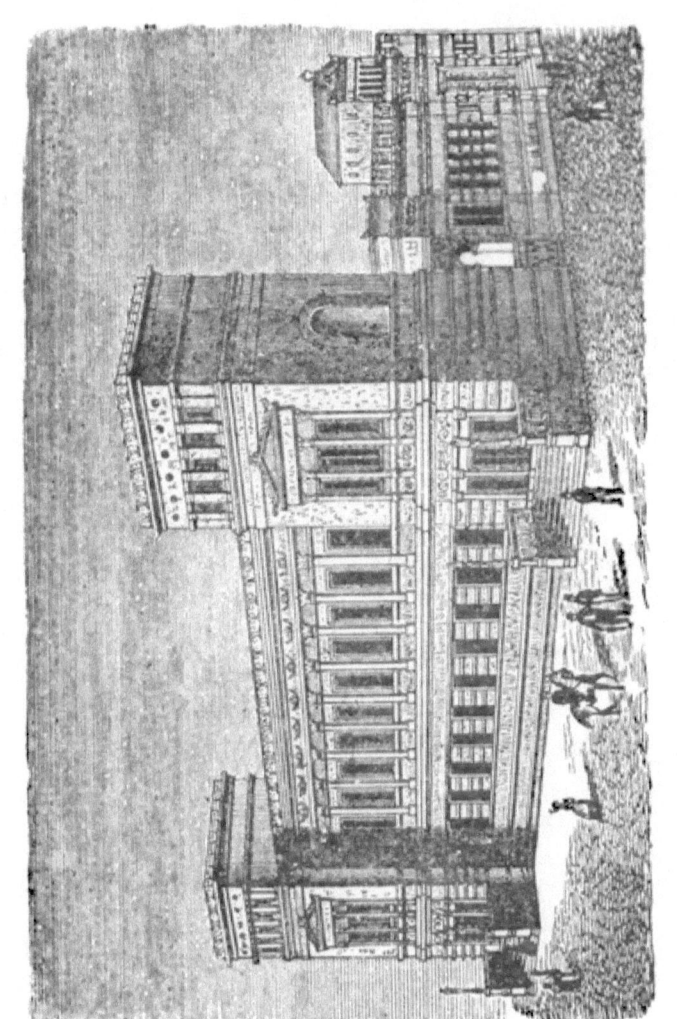

FIG. 18. — Chemical laboratory building used by Professor Kekulé of the University of Bonn.

to its neighboring atoms, on the one hand by two points and on the other hand by one point of attraction.

Recent studies of van 't Hoff, Pasteur, Wislicenus, and others have carried the subject still further. Attempting to explain certain remarkable peculiarities of carbon compounds, the theory has now been presented that the four points of attraction of a single atom of carbon are ordinarily distributed about its central point somewhat as are the four apexes of the regular tetrahedron. Again, when the series of carbon atoms are attached in circuit, it is supposed that in some cases one carbon atom is related to its neighbor as if the apex of a tetrahedron were presented to the apex of another tetrahedron. Again, in another case, a carbon atom may be attached to its neighbor as if an edge of a tetrahedron were presented to the edge of another tetrahedron. In the former case it is supposed that the one carbon atom is capable of a certain amount of rotation about the other carbon atom. In the latter case such rotation would naturally be impracticable. It is supposed that this theory contributes something toward explanation of asymetric carbon compounds; such, for instance, as the different varieties of tartaric acid.

CHAPTER IV.

THE THREE STATES OF MATTER.

SOLIDS, LIQUIDS, AND GASES.

The various kinds of matter known are capable of arrangement in a series representing many stages of tenuity. Between the most rigid solids on the one hand, and the most rarified gases on the other, there exist substances which possess many degrees of density, the progress from one extreme to the other being exceedingly gradual.

There are even no well-defined lines of demarcation between a solid and its own liquid, and between a liquid and its own vapor. Indeed, Thomas Andrews has shown that in case of the same substance there may be even a continuity of its liquid and gaseous form. Thus water may exist in a closed vessel under such conditions, at once of high temperature and pressure, that the upper portions shall be gaseous and the lower portions liquid, and yet the latter may show no *top surface* or other evidence of a distinct place where liquid ends and vapor begins.

For purposes of convenience, however, chemists admit the existence of three, and perhaps four, states of matter. They are called respectively the solid, the liquid, the gaseous, and the radiant forms.

From what has just been said, however, it is evident that it is not easy to define the four terms used.

A Solid Defined. — In general, a solid is a form of matter possessing such rigidity that a given portion of it resists with considerable vigor any forces that tend to change its shape. Of course, then, a given solid does not — unless under very great pressure — assume the shape of a vessel in which it is placed, and except, of course, the one was contrived to fit the other, or the solid is in the form of powder. (Possibly the latter qualification is unnecessary.)

A Liquid Defined. — In general, a liquid is a fluid form of matter having such internal mobility as enables it to assume the form of at least the lower portions of its containing vessel, whatever the shape, and provided only that the liquid shall be in sufficient quantity. Further, an untrammelled liquid has usually a horizontal top surface.

A Gas Defined. — In general, a gas is a fluid form of matter such as readily (under ordinarily prevailing conditions) assumes the shape of the containing walls of any vessel whatsoever. Far from having any distinct *top surface* of its own, a gas has sufficient expansive force to lead it to fill entirely any containing vessel whatever its size.

One of the most characteristic features of bodies when in the gaseous state is the great change in volume which they may be easily made to undergo. Their volumes are particularly sensitive to change of pressure and of temperature. When the gas is heated, great expansion takes place if the walls of the vessel are such as permit it. (If expansion is fully resisted, however, there is still an increase of tension or expansive force imparted to the gas. Hence addition of heat may produce increase of volume of a gas when pressure is maintained constant, or it may produce increase of tension when volume is maintained constant.)

Andrews recommends that the term *gas* be reserved for a substance when it is at a temperature higher than its critical point. He recommends that the term *vapor* be applied to a substance, — in the gaseous condition, — but existing at a temperature below its critical point.

The term *vapor* is applied, in a general way, to those gases which are easily condensible to the liquid form.

The modern scientist accepts the hypothesis that even a given small portion of gaseous matter is made up of millions of millions of molecules, and that these are in rapid motion in all directions. In all ordinary gases the length of the mean free path of a molecule is exceedingly small as compared with the dimensions of the confining vessel; in fact, in every second of time each molecule has millions of collisions with other molecules. Indeed, many of the well-recognized properties of gaseous matter are referable to these constant encounters.

Importance of the Study of Gases. — Several of the general laws of matter relate to it while in its gaseous condition. They show that bodies of the most diverse chemical constitution and properties, possess many close physical correspondences when in the gaseous form, while the solid and liquid forms possess no equivalent resemblances.

These facts create the belief that there is some fundamental similarity and simplicity of *molecular* condition characterizing the gaseous state. Hence this condition has been very carefully studied as that affording to molecular science the best common ground for the comparison of bodies. The hypothesis of Avogadro and Ampère is a result of this view. (See p. 78.)

It is probable, however, that a given substance may possess in the *solid* condition and in the *liquid* condition molecules containing a greater number of atoms than it has while in the *gaseous* condition.

Thus it has been shown by Raoult and by Thomsen, from very dissimilar points of view, that the molecular formula of *liquid water* is probably as great as H_4O_2. (This does not invalidate the statement of the formula of *water vapor* as H_2O.)

It is probable that the systems of molecules in liquids and solids are not only very different from those of gases, but that they vary greatly among themselves. These molecular aggregations, too, are probably not permanent, but are continually breaking up, their constituents changing partners.

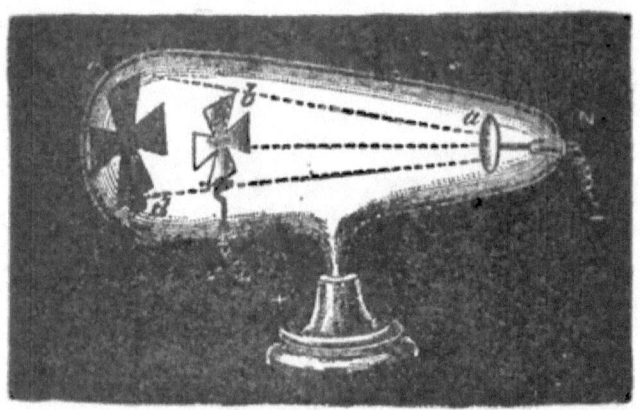

FIG. 19. — Crookes's radiant matter bulb. At *b* is a screen. When the electric current traverses the radiant matter in the bulb, the molecules fall against the screen *b*; thereupon they are intercepted, and a dark shadow is thrown at *d*.

Radiant Matter. — The radiant form of matter is much more tenuous than even the gaseous, which it most resembles. In the radiant form the molecules are very far apart compared with their distances in the gaseous form. It is this circumstance which leads them to display physical properties different from those of the same bodies in the gaseous condition. As a result, also, the mean free path is rendered so long that the molecular hits in a given time may be disregarded in comparison with the misses; thus each molecule is allowed more fully to obey its own laws without interference. When the free path becomes so long as to be comparable with the dimensions of the vessel, the properties which characterize the gaseous condition are reduced to a minimum,

FIG. 20. — Appearance of the illuminated screen and the dark image formed in Crookes's bulb, shown in Figure 13.

and the matter becomes exalted to the ultra-gaseous state called the radiant condition.

The study of radiant matter has been conducted of late years with great earnestness and success by Pro-

FIG. 21. — Crookes's bulb. The bulb contains a small amount of gas in the radiant form. When the electric current is passed through the tube, the molecules of gas strike against the paddles, setting the wheel in motion and making it traverse the track from one end of the bulb to the other.

fessor William Crookes of London. His experiments have been conducted in glass tubes or bulbs from whose interiors the major portion of gas originally within them has been removed. The minute portions remaining are then in the radiant form. Professor Crookes has investigated the properties of radiant matter by the use of these tubes and certain ingenious mechanical appliances constructed within them. He seems to have demon-

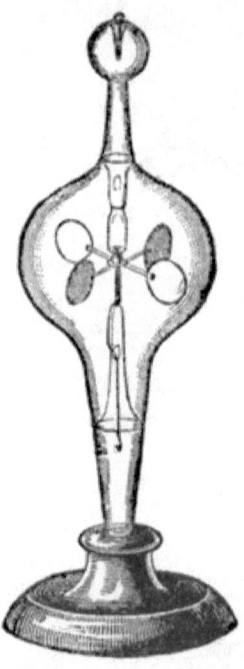

FIG. 22. — Crookes's radiometer. The bulb contains a very small amount of gas in the radiant form. Under the influence of a very minute amount of heat, the radiant matter causes the vanes of the mill to rotate rapidly.

strated a number of propositions like the following: (1) The length of the mean free path of molecules of matter in the radiant condition can be measured; (2) radiant matter exerts powerful phosphorogenic action where it strikes; (3) radiant matter proceeds in straight lines — it will not turn a corner; (4) radiant matter when intercepted by solid matter casts a shadow; (5) radiant matter exerts strong mechanical action where it strikes — this is proved by various forms of the well-

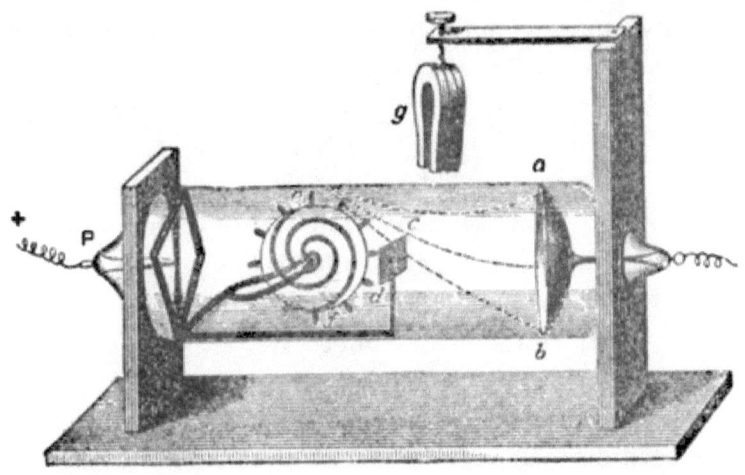

FIG. 23. — Crookes's bulb. The bulb contains a small amount of gas in the radiant form. Under ordinary conditions, even when the electric current is flowing through the bulb, the gas is intercepted by the screen cd. When, however, the magnet g is placed as indicated, the molecules are attracted so that they flow over the screen. They strike the paddles of the mill-wheel, setting it in motion.

known apparatus called Crookes's radiometer; (6) radiant matter is deflected by a magnet; (7) radiant matter produces heat when its motion is arrested.

CHAPTER V.

CHANGE FROM ONE STATE TO ANOTHER.

INFLUENCE OF ADDITION AND OF WITHDRAWAL OF HEAT.

Introduction. — Of the seventy chemical elements, five are gaseous at ordinary temperatures, — hydrogen, nitrogen, oxygen, fluorine, and chlorine. Two are liquids, — bromine and mercury. The others are solids.

The terrestrial globe (with its oceans and atmosphere) contains a very large number of compound substances. Under the conditions of temperature prevailing, the majority of them are *solid;* a considerable number exist in the atmosphere in the *gaseous* form; there is one widespread *liquid,* — water. (The water of the ocean, it is true, holds in solution a good many substances which may be considered to exist — for the time being — in the liquid form.)

Chemists have recognized an enormous number of compound substances other than those existing in nature. Of these, some are solid at ordinary temperatures, some are liquid at ordinary temperatures, some are gaseous at ordinary temperatures.

Influence of Heat. — In general, the condition of a substance, whether solid, liquid, or gaseous, is mainly dependent upon the amount of heat it contains. Of course the foregoing discussion has dealt with substances under *ordinary conditions* of temperature and under the possession by each individual substance of such an amount of heat as they may take up from the terrestrial system of earth, water, and air.

But chemists are able to add to or subtract from substances a very large amount of heat as compared with what the substances absorb from their natural surrounding medium. Incidentally the temperature of the sub-

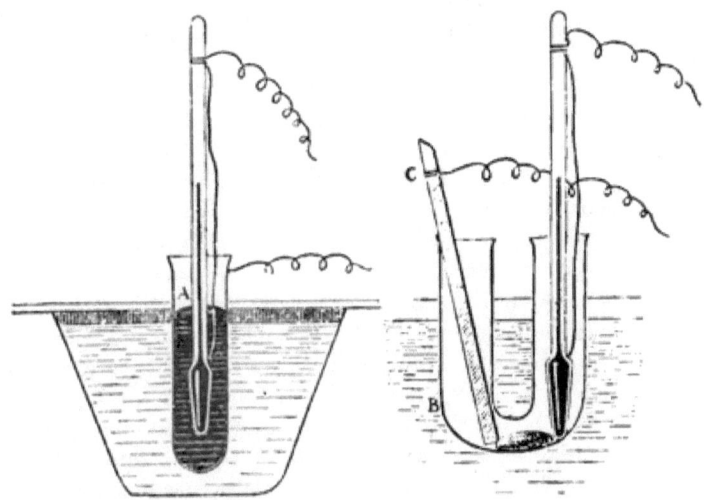

FIG. 24. — Disposition of apparatus for determining the melting-point of a solid. At the moment the solid liquefies electric communication is established through the wires and a bell is rung. At the moment of ringing the temperature is read off from the thermometer.

stance is raised or lowered by addition or withdrawal of heat respectively, although not in exact proportion to the amount added or withdrawn. The actual change of temperature is modified somewhat by certain natural properties of the substance, which give rise to the phenomena of specific heat. (See p. 45.)

The addition of heat to a body tends also to advance it from the solid to the liquid, and even to the gaseous state. And likewise the withdrawal of heat from a body tends to reduce it from the gaseous to the liquid and even to the solid state. The word *tends* is used here with the admission that, while in some cases the tendency is realized and the result sought is attained, in others it is not. In a few of the latter cases the requisite amounts of heating or cooling influences are not obtained by means at present at the command of the chemist. In many cases dissociation interferes. (See p. 49.)

A Double Change of State. — Certain well-known substances may be made at will, by addition or subtraction of moderate quantities of heat, to occupy either the solid, the liquid, or a gaseous state. Sulphur and phosphorus are examples of such elementary substances, and water is an example of such a compound substance.

It is not easy, however, to give *many* familiar examples of this sort.

A Single Change of State. — There are, however, many well-known cases of substances ordinarily solid that may be changed to the liquid state.

Thus, probably every one of the chemical *elements*, solid at ordinary temperatures, may be liquefied (carbon, perhaps, excepted). Again, the two liquid elements, bromine and mercury, may be vaporized. Further, the gaseous elements may all be liquefied and probably even solidified.

Most *compound* substances, whether natural or artificial, which are solid at ordinary temperatures of the earth, may be liquefied; and in most cases these liquids — and also compound substances liquid at ordinary temperatures — may be vaporized. An important exception must of course be made of that large class of organic substances (especially the organized ones), which, ordinarily solid, undergo dissociation — and that without liquefying — by addition of a moderate amount of heat. The same statement applies also to certain substances which may become liquid, but dissociate upon heating, before changing to the state of vapor.

In general, however, the chemist assumes that there are conditions — perhaps not always easy to secure — under which every substance, with the general exceptions noted, may be made to assume the solid, the liquid, or the gaseous state.

CHAPTER VI.

CHANGES INCIDENT TO ADDITION OF HEAT.

ADDITION OF HEAT TO A SOLID.

When heat is added to a solid, a series of effects may be produced in more or less orderly sequence, somewhat as follows: (1) rise of temperature according to the specific heat of the substance; (2) melting, associated with disappearance of heat under the conditions called latent heat; (3) a series of effects upon the liquid as, for example, rise of temperature according to the specific heat of the liquid; (4) partial vaporization followed by boiling of the liquid; (5) complete change of a liquid to the form of vapor; (6) incidental to many of these stages is expansion, whether of the solid, the liquid, or the gas; (7) dissociation may take place at different points of temperature according to the nature of the substance, *i.e.* in some cases a solid may dissociate, in other cases a liquid may dissociate, and in yet others dissociation may only take place after the substance has attained the gaseous form and the gas has been subjected to an extremely large addition of heat; (8) finally evolution of light may occur.

Rise of Temperature according to Specific Heat. — The specific heat of a substance is the special amount of heat involved in its rise or fall of temperature. To change a pound of water one degree upward in temperature, a large amount of heat must be added to it. To change a pound of water one degree downward in temperature, a large amount of heat must be subtracted from it. Most other substances have lower specific heats than water; *i.e.* a smaller amount of heat, added or subtracted, changes the temperature of a pound of any substance other than water by one degree upward or downward respectively.

TABLE OF A FEW SPECIFIC HEATS.

	Atomic Weight. Hydrogen = 1.	Specific Heat. Water = 1.
Sodium	23.	.293
Potassium	39.	.166
Iron	55.9	.112
Copper	63.2	.095
Bromine (solid)	79.8	.084
Silver	107.7	.057
Iodine	126.6	.054
Platinum	194.4	.032
Mercury (liquid)	199.7	.033
Mercury (solid)	199.7	.032
Uranium	238.5	.028

Gradual Melting. — In some cases of addition of heat to a solid, liquefaction seems to be a gradual process, and the temperature of the solid steadily rises. Sealing-wax illustrates this case. It becomes softer and softer until distinctly liquid. This kind of melting is most common in substances of decidedly heterogeneous composition.

CHANGES INCIDENT TO ADDITION OF HEAT. 47

Partial Melting. — In other cases addition of heat fully melts a portion of the mass of solid, leaving the rest hard. Little by little the whole liquefies.

This kind of melting is most common in bodies of uniform structure and marked crystalline tendencies.

Disappearance of Heat during Liquefaction (Latent Heat of Liquefaction). — A most important feature in these latter cases is the fact that, from the beginning of liquefaction to its completion, *there is no rise of temperature.* Large amounts of heat may thus be rendered *latent* without effect upon the *temperature* of the mass. Ice is an example of this process.

When ice of the temperature of $0°$ C. receives such an addition of heat as will but just liquefy it, the water so produced does not rise in temperature, but still remains at $0°$. In fact, one kilogramme of ice absorbs, while melting, 79 units of heat without any rise of temperature at all. The heat so absorbed is called latent heat. That it has merely undergone some temporary modification appears from the fact that when water is solidified into ice it always gives out 79 units of heat; in other words, the precise quantity it previously disposed of. Moreover, the water makes this latter change without any fall of its temperature.

In cases like the foregoing, the heat which does not increase the temperature is supposed to do some kind of internal work in changing the structure or positions of molecules.

In general, a definite amount of heat becomes latent when any solid is changed to the liquid form, and it reappears when the liquid changes back again to the solid form.

That solids in changing to the liquid form do indeed absorb heat is easily proved in cases like that of snow or ice melting in the hand.

Exceptional Cases. — There are certain exceptional cases in which a solid, when heated, turns so quickly to the gaseous form that it seems as if it did not exist in the liquid form at all. Solid iodine is an illustration.

Experiment. — Heat about one gramme of iodine crystals in a large flask, with the intention of liquefying it if possible. It may be noted that the substance turns quickly to gas, apparently without liquefying at all.

Observe also that the iodine vapor condenses in the upper part of the flask immediately to the solid form. If it condensed to the liquid form, it might be expected to form in drops or run down in streams, as many other condensing gases and vapors do.

NOTE. Remove the iodine from the flask as follows: Introduce a few c.c. of water and about one-half gramme of solid potassic iodide. Let the solution flow to different parts of the flask so as to dissolve the iodine.

Spontaneous Melting in the Air. — Certain solids melt upon exposure to air of the ordinary temperature. Ice is an excellent example. It melts by the addition of heat from the atmosphere or from any surrounding solid or liquid bodies. The surrounding substances are of course cooled by the process. Many cooling operations in the arts depend upon this principle. It must be observed, however, that the heat lost by the cooled objects is absorbed by the melting ice as latent heat.

Special Forms of Liquefaction. — In each of the cases, previously referred to, a single individual substance has been assumed to liquefy by the addition of heat. There are many cases in which two substances together form a single liquid. Thus two solids, when properly intermingled, may change to a liquid. For example, solid ice and solid salt, when pulverized together, turn into a liquid. Then the one substance dissolves in the other; *i.e.* the solid salt dissolves in the liquid water.

In some cases a solid becomes liquid by solution in some liquid. Thus liquid water and solid common salt, when intermingled, may form a single liquid. Of course a solid salt has dissolved in the liquid water.

Thus ammonic nitrate, ammonic acetate, potassic iodide, and many other solids, when placed in a small amount of water, dissolve very rapidly; at the same time they absorb heat to a most marked degree.

In all such cases heat is absorbed. It may be called latent heat of solution as distinguished from latent heat of liquefaction already referred to.

Other Special Cases. — Many special cases are known in which a mixture of two or more solids melts at a lower temperature than that at which the *easiest-melting one* melts alone.

Thus certain alloys of bismuth, tin, lead, and cadmium, called fusible alloys, melt at the temperature of about 160° Fahrenheit; that is, fifty degrees below the boiling-point of water. But the melting-point of tin alone, the easiest melting, is 451° F.[1]

So also certain mixtures of chemical salts, called salt alloys, illustrate the same principle by melting at a lower point than either salt singly.

These and other somewhat similar phenomena are classed under the term *eutexia*. An eutectic mixture of substances is that mixture which possesses a lower liquefying-point than that of either constituent separately or any other mixture (not chemical compound) of them.

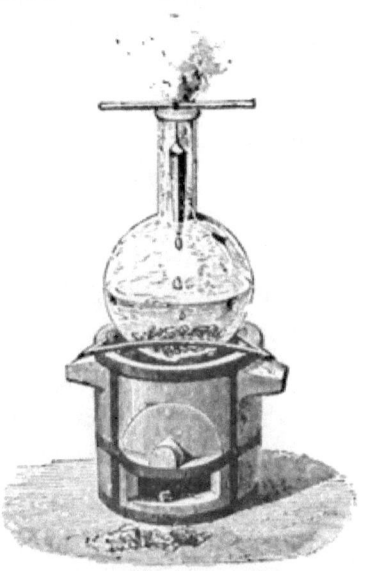

FIG. 25. — A rod of fusible alloy is suspended in the steam rising from boiling water. The rod melts and falls in liquid drops.

These phenomena must probably be referred to some obscure chemical combination of the substances involved. (See p. 172.)

Dissociation. — In case of certain substances dissociation takes place at a temperature far below that at which

---

[1] Melting-point of tin alone . . . . 451° F.
  "   "   " bismuth alone . . . . 515° F.
  "   "   " cadmium alone . . . . 608° F.
  "   "   " lead alone . . . . 619° F.

the substances become luminous. The familiar case of mercuric oxide affords an example. On the other hand, some substances seem never to be decomposed by mere accession of any such quantity of heat as we can add to them. Silicic oxide, sand, is an example.

In a third class of cases dissociation occurs, but it is merely temporary; that is, if immediately after dissociation the temperature of the substances falls slightly, the elements recombine so that the original compound reappears as if it had never been decomposed at all. For this reason many of the facts of dissociation long escaped notice. Of late, however, new methods of experimenting have made it possible to prove dissociation in cases in which it was at first unsuspected, and later even denied.

When dissociation does occur, it may be supposed that the accession of heat has become so great as to produce not only motion of the molecules as wholes, but, further, to produce a motion of the atoms within the molecules; in its earlier stages producing some of the phenomena of specific heat, in its later ones the heat seems to increase the atomic motion until the constituent atoms move through such distances as throw them outside of the range of the force of chemical affinity. Then occurs dissociation.

It may be added parenthetically that in the view of certain eminent investigators this line of research suggests the probability that what are commonly regarded as ultimate atoms may themselves be dissociated by a still greater heat. In this view the number of true elementary substances is much smaller than is now admitted. (See p. 29.)

ADDITION OF HEAT TO A LIQUID.

When heat is added to a liquid, a series of effects may be produced in more or less orderly sequence; they correspond tolerably well to those associated with the addition of heat to a solid. They are somewhat as follows: (1) rise of temperature; (2) expansion; (3) vaporization, associated with disappearance of heat as latent heat of vaporization; (4) boiling; (5) dissociation may or may not occur; (6) if the addition of heat is continued, evolution of light may be produced.

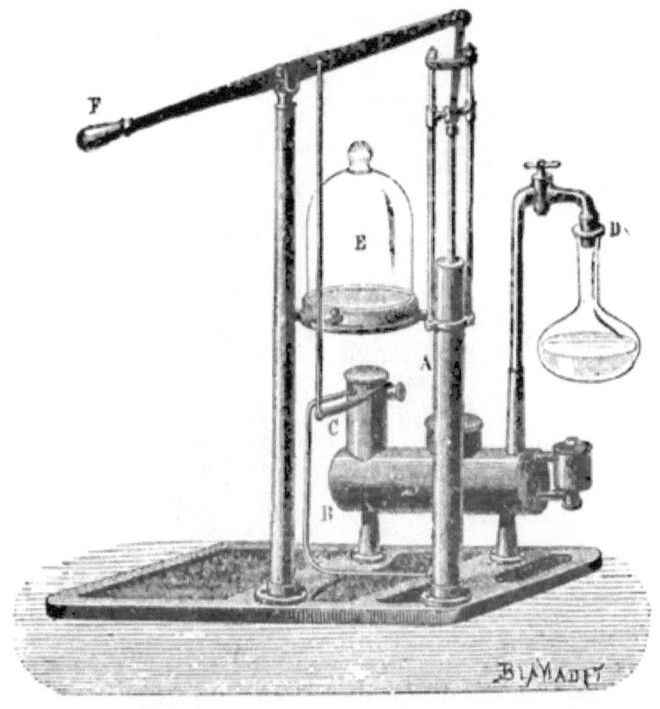

FIG. 26. — One form of Carré's ice machine The purpose is to freeze the water in the carafe *D*. When the air-pump is set in motion, the air, and at the same time the water vapor, is withdrawn from the flask *D*. The water vapor is absorbed by concentrated sulphuric acid placed in the reservoir *B*. This absorption is rendered the more prompt because the sulphuric acid is agitated by a plunger in *C*. The very rapid evaporation from the surface of the water in *D* absorbs so much latent heat from the water as to freeze it.

Rise of Temperature. — When heat is added to a liquid, the temperature rises, according to the relations of the liquid to specific heat, until the boiling temperature is reached. The boiling-point varies not only for different liquids, but varies for the same liquid according to the pressure at the time prevailing.

Vaporization of Liquids. — At all ordinary temperatures a liquid gives off vapor into the space above it. The amount of this vapor depends on the nature of

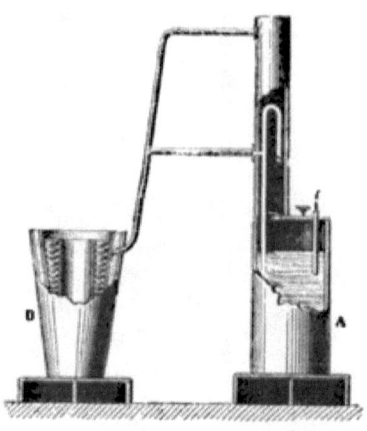

Fig. 27. — Small form of Carré's ice machine. Solution of ammonia in water is contained in the receiver A. Under the influence of heat, ammonia gas is driven into the jacket in B. The ammonia gas in this jacket is liquefied by reason of its own pressure. In the second stage of the operation the heat is withdrawn from A, whereupon the liquid ammonia in the jacket volatilizes rapidly. It returns to the water in A. By reason of its rapid evaporation, it cools that portion of water in the inner cylinder in B. The water is thereby frozen.

the liquid and the temperature prevailing at the time of the experiment. The vapor exerts a pressure called *vapor pressure*. At first vapor comes from the surface only. As the temperature rises, more and more vapor rises. At length the entire mass of liquid reaches a point at which bubbles of its vapor may form in parts of its interior, and may even acquire sufficient vapor pressure to come to the surface of the liquid without condensation. Then boiling begins.

During the continuance of boiling, large amounts of heat may be added to the liquid without raising its temperature above the *boiling-point for the particular pressure prevailing*. Such heat is called *latent heat of vaporization*. It is definite in amount. It is believed to do some kind of

FIG. 28. — Machine for producing artificial ice by the rapid evaporation of liquefied ammonia gas (NH$_3$).

internal work on the molecules of liquid in changing their structure or positions.

A marked concealment of heat takes place when water at 100° C. is changed into steam of 100° C. In this case also an amount of heat, equal to that made latent, reappears when the reverse change of state takes place. That is, one kilogramme of water of 100° C. in changing to vapor of 100° C. absorbs 536 units of heat (without any rise of temperature). Again, one kilogramme of steam at 100° C. upon condensing to one kilogramme of water at 100° C. evolves 536 units of heat.

In general, a definite amount of heat becomes latent when any liquid changes to the gaseous form, and it reappears when the gas changes back again to the liquid form.

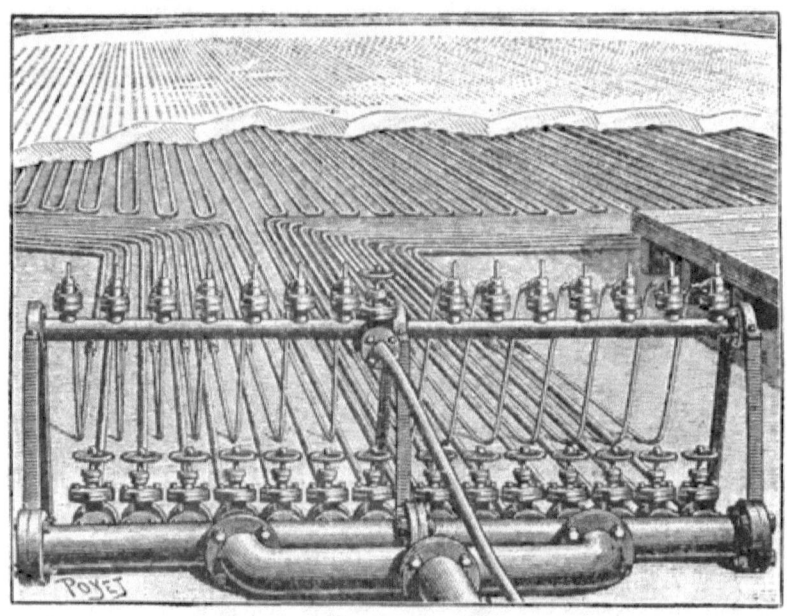

FIG. 29. — Method of employing currents of cold brine in circulating tubes for the purpose of freezing the surface of an artificial lake. The brine is cooled below the freezing-point of water by an ammonia ice machine.

Experimental Demonstration of Absorption of Heat during Vaporization. — That liquids do in fact absorb heat in vaporization is easily proved in case of such liquids as water, petroleum naphtha, ethyl ether, etc.

Other examples are found in the evolution of a gas from its dissolving liquid, as when ammonia gas escapes from a concentrated water solution of ammonia.

Spontaneous Evaporation. — Certain substances evaporate very readily when exposed in open vessels to the ordinary temperature of the atmosphere. In other

FIG. 30. — Skating-rink in Paris. The ice is produced by a circulation of cold brine, as shown in Figure 29.

words, they have low boiling-points. Ethyl ether (also called sulphuric ether) is a good example of a liquid of this sort. Of course in this, as in all cases, heat is rendered latent. Heat is absorbed from neighboring objects, and, as the evaporation proceeds rapidly, rapid absorption of heat is associated with it. The ordinary statement is that such evaporation precedes cooling.

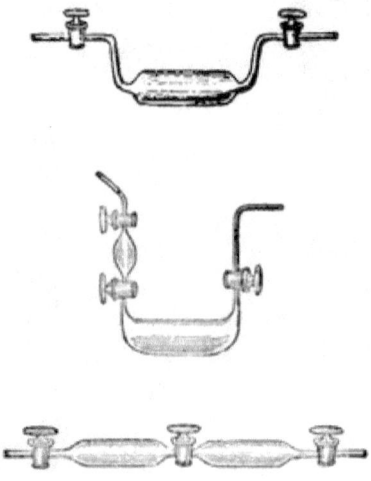

FIG. 31. — Three forms of receivers for collecting sulphur dioxide changed from the gaseous form to the liquid by cold and pressure. The second and third forms are so constructed that a small amount of liquid may be transferred from one bulb to another, and then may be withdrawn for experiment.

Ice Machines. — The ice machines in common use depend upon the principles already stated. The substance oftenest used is ammonia gas. In one part of the machine ammonia gas is condensed to a liquid by the aid of pressure and a *considerable amount* of cool water circulating about the vessel in which the ammonia gas is confined.

Next the liquid ammonia is transferred to another portion of the apparatus. Here its vessel is in indirect contact with the water to be frozen. Upon opening a stopcock, the liquid ammonia rushes into vapor, instantly absorbs an immense amount of heat from the water, and thereupon solidifies it.

CHANGES INCIDENT TO WITHDRAWAL OF HEAT.

WITHDRAWAL OF HEAT FROM A GAS.

When heat is withdrawn from a gas or vapor, the substance manifests various phenomena, similar to those already referred to under the head of addition of heat, only, of course, in the reverse order.

They are, among others, the following: (1) Cessation of the evolution of light; (2) recombination (in some cases of dissociation); (3) fall of temperature; (4) diminution of volume; (5) change to the liquid form; (6) evolution of latent heat.

If the liquid formed is subjected to further withdrawal of heat, other changes follow. (See p. 59.)

Liquefaction of Gases. Latent Heat. — In accordance with the general principles already enunciated, it may be expected that liquefaction of gases should be attended with evolution of heat. This is indeed the case. In changing from the state of gas to the state of liquid, heat is given out. The heat is absorbed by any suitable body. If the body is cold, it becomes warm thereby; if the body is warm, it becomes warmer still.

These two statements, simple as they appear, are worthy of further examination. A cold body condenses vapors, the cold body thereby becoming slightly warmer, of course. Almost the only commonly-known case of a moderately warm body condensing vapor is where some warm body condenses steam; *i.e.* water-vapor. If steam falls upon a person's hand, the steam condenses to drops of water. At the same time latent heat is evolved, and the hand may be severely burned. Again, in steam-heating appliances the steam, condensing in the suitable pipes, evolves its latent heat by the act of condensation, and thus the air of the building may be warmed.

58 CHANGES INCIDENT TO WITHDRAWAL OF HEAT.

Influence of Pressure. — Of course the liquefaction of gases is easier under increase of pressure; but, after all, *pressure* is only a secondary consideration.

The English scientist, Thomas Andrews, discovered that however great the mechanical pressure upon a gas, this pressure cannot reduce the gas

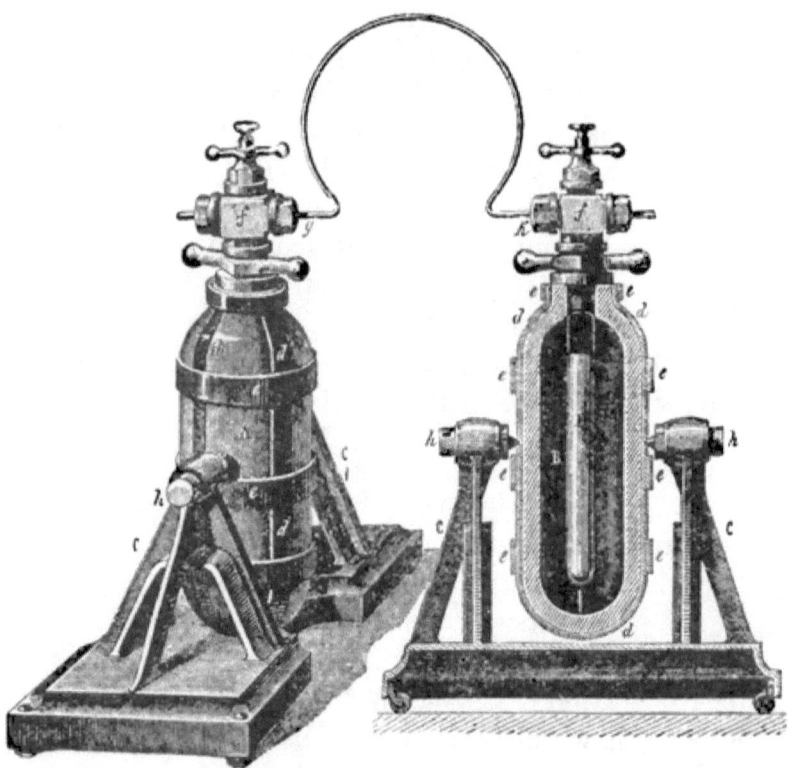

FIG. 32. — Deleuil's apparatus for the liquefaction of carbon dioxide gas. The gas is generated in the right-hand vessel. The inner tube, containing sulphuric acid, is broken by forcing the tube against the pin at the bottom of the receiver. The acid acts upon carbonate of soda, liberating carbon dioxide gas. The latter passes into the left-hand receiver, and is there condensed into the liquid form by influence of the great pressure of gas.

to the form of liquid, *except when the gas is at or below a certain point of temperature*. This point differs for different gases. It is called in each case *the critical point*.

The same facts may be stated, in a sort of reversed form, as follows: for each liquid *there is a point of temperature* at which this liquid will assume the gaseous form, no matter how great the mechanical pressure then opposing it. This is another definition of the critical point.

WITHDRAWAL OF HEAT FROM A LIQUID.

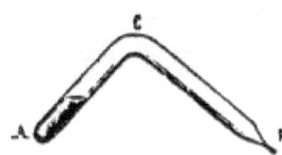

FIG. 33.—Tube showing liquid ammonia (NH₃) at *A*. Part of the ammonia gas in the tube is liquefied by great pressure and cold.

When heat is withdrawn from a liquid, a series of effects are manifested as already indicated. They are in the reversed order of those noted when heat is added to a solid.

If the withdrawal of heat continues, the liquid begins to fall in temperature according to its specific heat. In due time the liquid may change to a solid. For the moment no fall of temperature takes place, but the heat withdrawn from the liquid is supplied by the liberation of latent heat. As soon as solidification has become complete the evolution of latent heat ceases, and the solid begins to fall in temperature in accordance with its specific heat. At the same time, of course, it contracts in volume. The general statement of the series of phenomena associated with withdrawal of heat is now complete.

FIG. 34.—Disposition of apparatus for liquefying ammonia gas. In the heated water-bath, solution of ammonia gas in water is decomposed. The gas passes into the other branch of the tube, which is surrounded by ice. Under the influence of the cold, and the great pressure of the ammonia gas, a part of the latter is liquefied.

Solidification of Homogeneous Liquids. — A liquid consisting of a single elementary substance, or a liquid consisting of a single compound substance, may be called, in a general way, homogeneous. (See p. 17.) When such liquids solidify, they may, under certain conditions, crystallize; under others, not. The size of the crystals, and in some cases the particular form, depends upon the circumstances under which the solidification has taken place.

Solidification of Mixed Liquids. — There are many cases known in which two distinct solids, liquefied by natural or artificial high temperature, are capable of

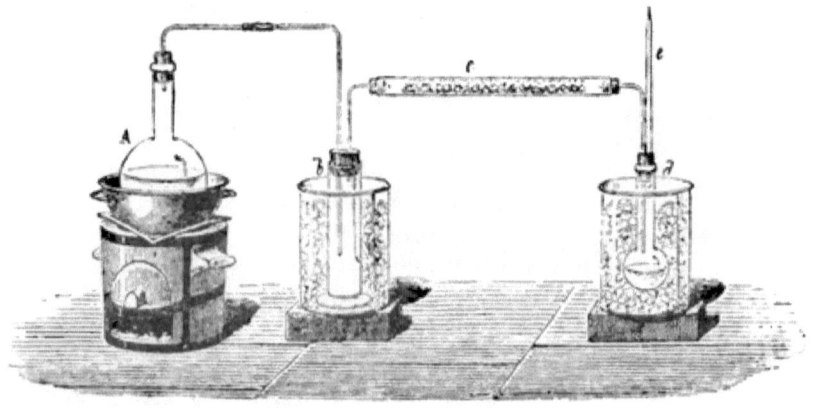

FIG. 35. — Apparatus for liquefying sulphur dioxide gas. The gas generated in *A* is purified in the wash-bottle *b* and the drying-tube *c*. It then condenses to a liquid in the flask *d*, which is surrounded by a freezing mixture.

very complete liquid interdiffusion. They thus form what may be regarded as a single liquid — so long as the temperature does not fall the constituents do not separate from each other. Of course, however, a sufficient withdrawal of heat produces solidification. It has been observed, further, that when the cooling proceeds *very gradually*, a series of different substances separate progressively from the liquid. The usual course is as follows: first, one of the component solids separates;

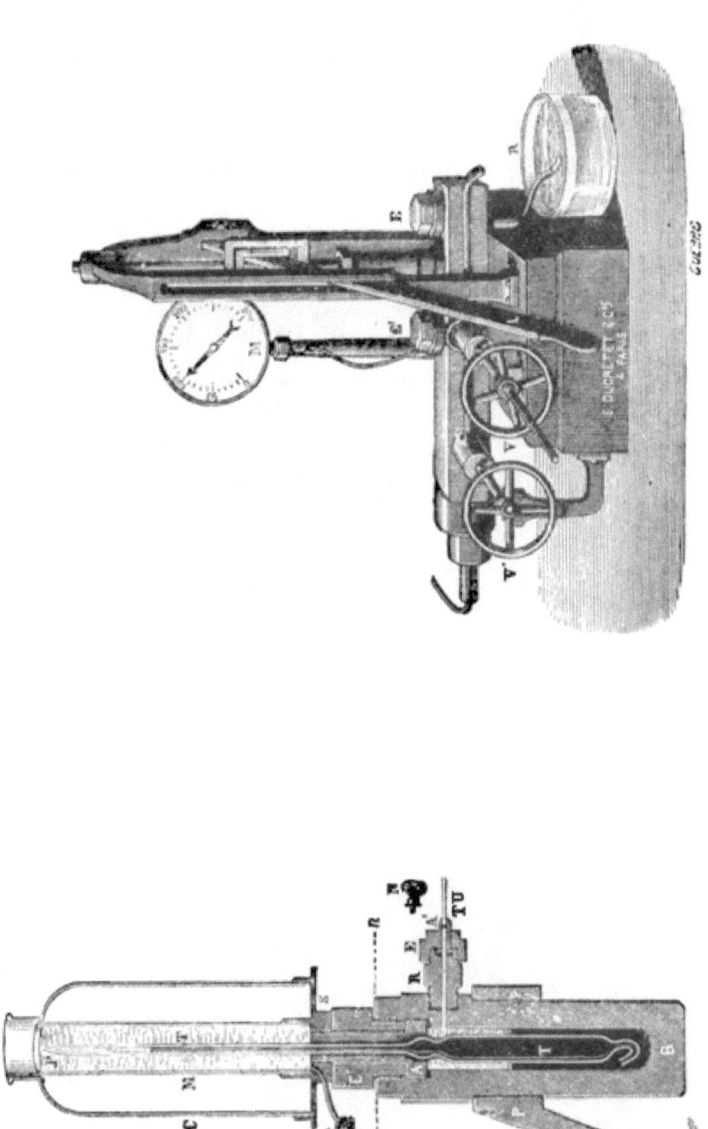

FIG. 36. — Cailletet's apparatus for the liquefaction of gases. It consists of the hydraulic pump, which, by the intervention of the mercury in the vessel T, produces pressure upon the gas in the tube P. The tube P is surrounded by refrigerating liquid.

next, some well-defined compound of the components crystallizes out; sometimes the operation terminates by

FIG. 37. — Part of Cailletet's apparatus, showing its position when the tube *P* is filling with the gas to be experimented upon.

the solidification of a nearly pure mass of the other component.

CHAPTER VII.

CERTAIN GENERAL LAWS OF MATTER.

SIGNIFICANCE OF THE GASEOUS CONDITION.

DALTON's atomic theory, as now received, may be said to have its basis in certain fundamental laws of matter. The following are some of the most important of them: —

Boyle's or Mariotte's Law of the Pressure of Gases: —

The volume of any gas, when confined under a constant and relatively high temperature, varies inversely as the pressure to which it is exposed.

The law is easy of comprehension.

It states certain facts that are capable of direct experimental demonstration.

It is of wide application, for it applies to any gas whatsoever.

It has but one limitation. That is, a high temperature is necessary. What, then, is high temperature? Evidently the term is relative, and its application must be interpreted according to the special properties of the gas in question.

The guide to this interpretation must be found in what is called *the critical point for a gas*. (See p. 58.)

When a gas is subjected to varying pressure at temperatures *far above its critical point*, it conforms closely to the law as stated.

At temperatures *below its critical point* a gas is ready to change to a liquid by a slight increase of pressure. But this change brings the substance to a condition in

FIG. 38. — Robert Boyle. Born in 1626; died in 1691.

which it obeys the laws of liquids, and not laws like that of Boyle and Mariotte, which relates to gases exclusively.

At temperatures *slightly above the critical point* a gas is found to manifest the beginnings of the influences of the laws of liquids, and so it does not conform closely to the laws of gases.

To these explanations of the law the following statement may be added:—

If a given volume of any gas whatever is subjected to the single change of having the external pressure upon it doubled, the volume of the gas is thereby reduced to one-half; and the converse is likewise true — that is, if any gas is subjected to the single change of having the external pressure upon it reduced to one-half, then its volume thereby becomes double.

But a given gas strictly conforms to this law only when it is at temperatures far above that of its liquefying-point, wherever that may be.

Charles's Law of Expansion of Gases by Heat:—

The volume of a given portion of any gas, under a constant pressure, varies directly as the temperature expressed in absolute centigrade degrees.

Charles's law is a formal statement of facts that are capable of direct demonstration; it is one that is much used in the study of gases.

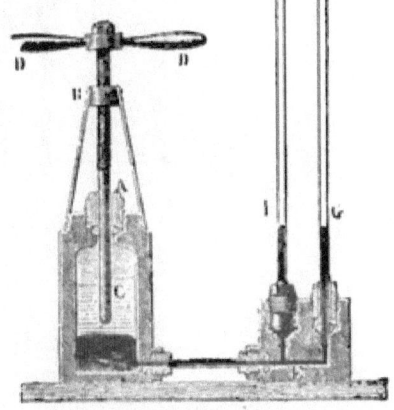

FIG. 39.—Apparatus for showing that different gases in the tubes I and G when subjected to the same amount of pressure, under condition of constant temperature, suffer the same amount of reduction of volume.

The ordinary centigrade thermometer reads 0° at the temperature of freezing water, and it reads 100° at the temperature of boiling water.

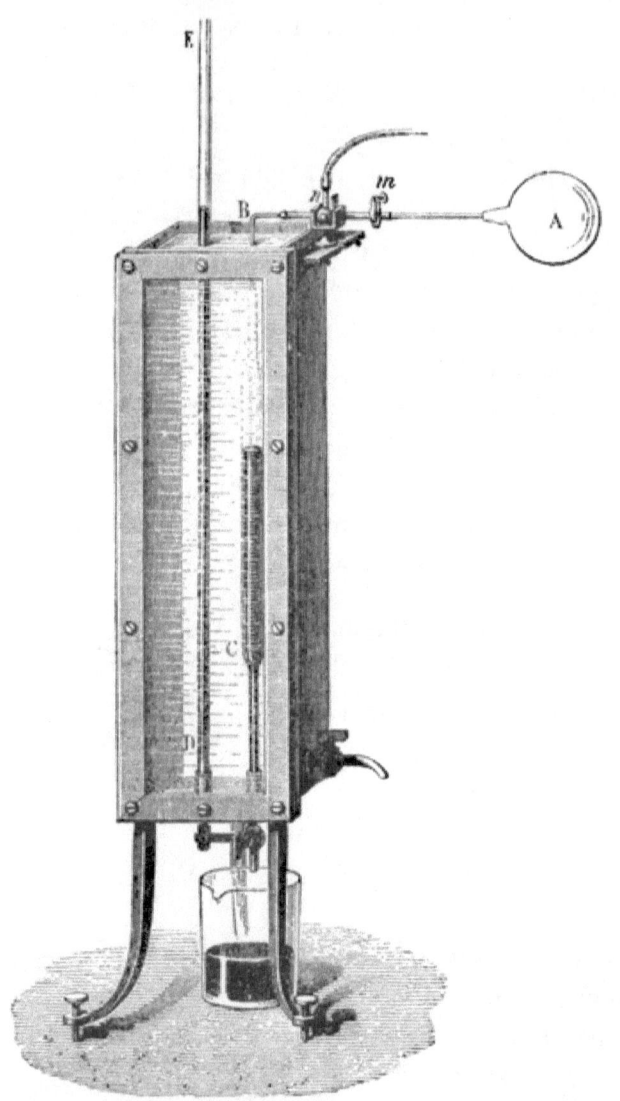

Fig. 40. — Apparatus for demonstrating the truth of the law of Charles. A portion of gas from the globe *A* may be transferred to the tube *BC*. Here it may be subjected to varying temperatures, from that of freezing water to that of boiling water. The change of volume of the gas may be read off either from the graduations on the tube *BC* or else by use of a cathetometer. At the same time the gas may be subjected to a given pressure by adding mercury to the tube *E*, or by withdrawing it from the bottom of the apparatus.

The absolute centigrade thermometer reads 273° at the temperature of freezing water, and it reads 373° at the temperature of boiling water.

Now the absolute centigrade scale is graded with reference to the fact that the coefficient of expansion of gases by heat is the same for all of them; namely, for an increase of one centigrade degree the expansion is $\frac{1}{273}$ of the volume the gas would occupy at the temperature of freezing water.

Since, then, the absolute scale is graduated at 273° at the freezing-point of water, the volume of any gas increases under the influence of added heat at the same rate as the numbers on the scale do.

Rule. Add 273° to any ordinary centigrade degree, and the sum gives the corresponding absolute centigrade degree.

Graham's Two Laws of Gaseous Diffusion: —

1. *The diffusion-rate of gases of the same density is the same, whatever their chemical composition.*
2. *The relative diffusion-rates of two gases of different densities are inversely as the square roots of these densities.*

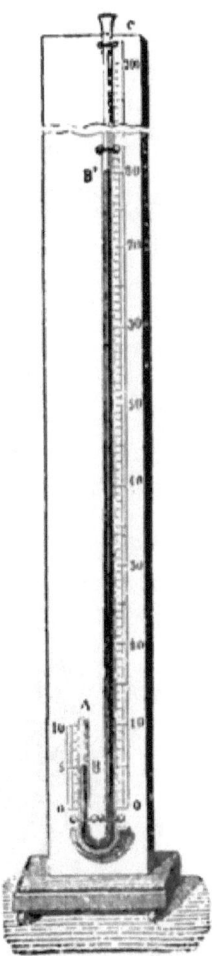

FIG. 41. — Apparatus for experimental proof of the law of Boyle and of Mariotte. Upon pouring additional mercury into the tube at *C*, additional pressure is applied to the portion of gas at *AB*. The volume of gas is thereby reduced in accordance with the law.

These laws are statements of facts capable of direct demonstration. When portions of gases of different densities are brought into direct contact, or are separated by a porous partition only, the molecules of both

FIG. 42. — Jacques Alexandre César Charles. Born in 1746; died in 1823. Celebrated French physicist.

gases are discovered to be in active motion. Incidentally the gases intermingle. These properties of gases are displayed by the various forms of the diffusiometer.

If a portion of gas is confined within the porous cell of a diffusiometer, this gas projects its molecules into

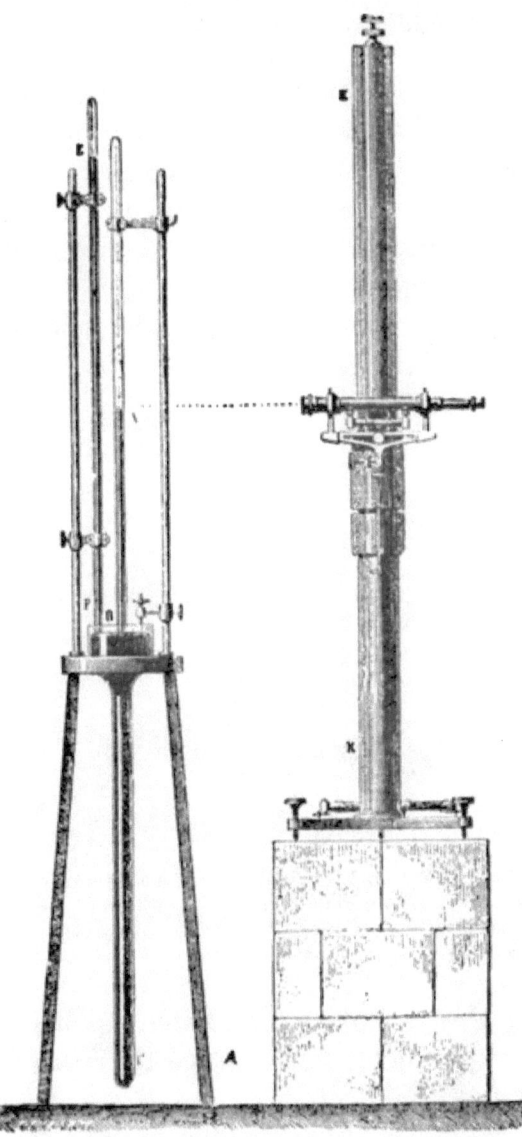

Fig. 43. — One form of Torricellian barometer, with its cathetometer. In this case the barometer is EF. The apparatus in the figure is intended to measure the pressure of mercury upon a portion of gas in the tube AB.

the passages of the porous walls, and they pass through and out. Simultaneously, and in the same manner, any external gas projects its molecules inward. But the rate of passage is found to be different if the gases have different densities. This is proved by the motion of the liquid in the gauge-tube connecting with the cell of the diffusiometer. It is observed that the molecules of the

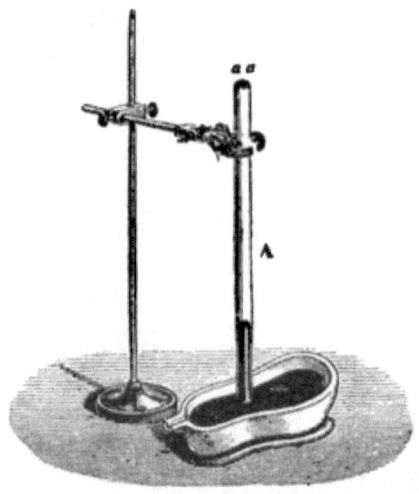

FIG. 44. — Graham's apparatus for showing diffusion of hydrogen gas. The tube *A* contains hydrogen gas. The top of the tube is closed by a porous wafer. The hydrogen gas escapes so rapidly into the air that the atmospheric pressure upon the mercury in the trough is able to force the mercury up into the tube *A*.

lighter gas always move with greater rapidity. A series of careful experiments has afforded the basis for the law already stated.

The following is an illustration of this law: A given bulk of oxygen gas is found by experiment to weigh sixteen times as much as the same volume of hydrogen gas. The density of oxygen is then said to be sixteen (it being customary to adopt hydrogen gas as the standard of density for gases). Now hydrogen gas is found experimentally to diffuse itself into oxygen gas four times as rapidly as oxygen gas diffuses into it.

The Law of Henry and of Dalton, of the Relation of Pressure to the Solubility of a Gas in Water:—

When a given gas is exposed to water under a constant temperature, the volume of the gas dissolved by the water varies directly as the pressure acting at the time.

Fig. 45.—Syphon for containing water saturated with carbon dioxide gas.

The amount of gas dissolved by water varies with the nature of the gas. It also varies with the temperature, being *in general* less at higher temperatures. It also varies with the pressure acting. This last is the only one of the three conditions that can be described in the form of a law. By the law as given, it appears that a gas resting on the surface of water is dissolved by the water to a certain extent. If now (other conditions being appropriate) the pressure, for example, is doubled, the amount of gas dissolved will also be doubled. It likewise follows that if the pressure is, for example, halved, the amount of gas dissolved will be halved; in other words, under lessened pressure gas dissolved in water actually comes out of the water, escaping with effervescence. If water, charged with gas under pressure, is allowed to flow from a *syphon*, the gas immediately leaves the water with effervescence.

The Law of Definite Proportions: —

The same compound always contains the same atoms, united in the same proportions by weight (and by volume when they are gaseous), and with the same molecular arrangement.

Example: Pure water always contains in each molecule only one atom of oxygen and only two atoms of hydrogen. It always contains these atoms in approximately the following proportions, by weight: —

    Hydrogen . . . 2 parts by weight.
    Oxygen . . . . . 16 parts by weight.
                                 18

It always contains these atoms in the following proportions, by volume: Two volumes of water vapor when decomposed yield approximately: —

    Hydrogen . . . . two volumes.
    Oxygen . . . . . . one volume.

The molecular arrangement is believed to be such that the oxygen atom is somehow between the two hydrogen atoms, — an arrangement which may be expressed by the formula, $H - O - H$.

This law contains several statements. Some of them are capable of experimental demonstrations; some of them are not. But the latter are based upon a multitude of observed facts which strongly suggest their truth.

The law as a whole is implicitly and safely relied upon in all chemical experiments and in the conduct of the great chemical industries of the world.

CERTAIN GENERAL LAWS OF MATTER.

The Two Laws of Multiple Proportions: —

1. *When the elementary substance A chemically unites with the elementary substance B in more than one proportion by weight (and in case of gaseous elements, by volume as well), they form more than one compound, and the several compound substances so produced possess well-marked and distinctly different properties.*

2. *The several amounts by weight (and in case of gaseous elementary substances, by volume also) of B, that may combine with the same amount of A, bear a very simple relation to each other.*

As a general example illustrating this law, suppose that the elementary substances A and B unite in the several proportions expressible by the formulas: —

AB_m, AB_n, AB_o, AB_p, AB_q. — It is found that the five compounds formed are essentially different substances. It is found that several amounts of B represented by m, n, o, p, q, bear very simple ratios to each other.

As a special example the compounds of nitrogen and oxygen may be taken.

These two substances unite in five different proportions by weight and gaseous volume.

They produce five distinctly different compounds, each having special characteristics of its own. They are the following: —

Nitrogen Monoxide (N_2O).

This is composed of 28 parts nitrogen by weight
and $\underline{16}$ " oxygen " "
44

Nitrogen Dioxide (N_2O_2 or NO).

This is composed of 28 parts nitrogen by weight
and $\underline{32}$ " oxygen " "
60

Nitrogen Trioxide (N_2O_3).

This is composed of 28 parts nitrogen by weight
and $\underline{48}$ " oxygen " "
76

Nitrogen Tetroxide (N_2O_4 or NO_2).

This is composed of 28 parts nitrogen by weight
and $\underline{64}$ " oxygen " "
92

Nitrogen Pentoxide (N_2O_5).

This is composed of 28 parts nitrogen by weight
and $\underline{80}$ " oxygen " "
108

Evidently in these five compounds the several weights of oxygen combined with the constant weight of nitrogen bear the simple ratios 1 : 2 : 3 : 4 : 5. (For further discussion of these compounds, see pp. 232, 236.)

NOTE. It will be observed that in the five compounds just cited in illustration of the laws under consideration, the amount of nitrogen is taken as a basis of comparison, and its weight is represented by the number 28. This number has been used because it has been found, after a great multitude of most carefully devised and executed experiments, that it has some special significance in this case. *It is believed to represent twenty-eight microcriths; i.e. the weight of the amount of nitrogen present in one molecule of each of the substances referred to.*

In its first stages quantitative chemical analysis makes its statements in percentage form. In the cases of the nitrogen compounds referred to, such statements are given in columns two and three of the following table: —

CERTAIN GENERAL LAWS OF MATTER.

1	2	3	4	5
	PER CENT BY WEIGHT.		WEIGHT-RATIO.	VOLUME-RATIO.
NAMES OF COMPOUNDS.	Nitrogen	Oxygen	Nitrogen : Oxygen	Nitrogen : Oxygen
Nitrogen monoxide	63.71	36.29	1 : .57 (1)	1 : $\frac{1}{2}$ (1)
Nitrogen dioxide	46.75	53.25	1 : 1.14 (2)	1 : 1 (2)
Nitrogen trioxide	36.91	63.09	1 : 1.71 (3)	1 : $1\frac{1}{2}$ (3)
Nitrogen tetroxide	30.50	69.50	1 : 2.28 (4)	1 : 2 (4)
Nitrogen pentoxide	25.98	74.02	1 : 2.85 (5)	1 : $2\frac{1}{2}$ (5)

If in these several compounds the amount by weight of nitrogen in each is taken as a common unit of comparison, then the amounts by weight of oxygen will be .57 : 1.14 : 1.71 : 2.28 : 2.85, and these numbers appear at once by inspection to be to each other as 1 : 2 : 3 : 4 : 5.

It is plain then that the several amounts of oxygen in these compounds, combined with the same amount of nitrogen, bear to each other the very simple ratio stated as 1 : 2 : 3 : 4 : 5.

In its most advanced stages, quantitative chemical analysis employs for each elementary substance a number representing its atomic or molecular weight. Thus the atomic weight of nitrogen is believed to be 14 microcriths, and its molecular weight 28 microcriths. The atomic weight of oxygen is believed to be 16 microcriths, and its molecular weight 32 microcriths.

Returning now to the table given, it is discovered that the volume-ratios afforded by the five compounds show yet more marked simplicity. Not only do the several volumes of oxygen bear to each other the simple ratios 1 : 2 : 3 : 4 : 5, but they also in each case bear a very simple ratio to the constant volume of the *other element*, the nitrogen. (See the laws of Gay-Lussac.)

Gay-Lussac's Three Laws of Combination of Gases: —

1. *When two or more gases combine, the volumes of these gases bear very simple ratios to each other.*

2. *When two or more gases combine to form a product which can remain a gas, the volume of the gas so formed*

bears a very simple ratio to the volume of each of the component gases.

3. *The weights of combining volumes of gaseous*

FIG. 46.—Gay-Lussac. Born 1778; died 1850.

elements bear very simple ratios to their atomic weights.

(In all these cases it is understood that, when under comparison, the gases are at constant temperatures and pressures.)

Many illustrations of the signification of these statements might be given. Two useful ones are presented here.

First Illustration drawn from Hydrochloric Acid Gas.
— The principal points to be mentioned in connection with the matter here under consideration may be expressed by the following equations:—

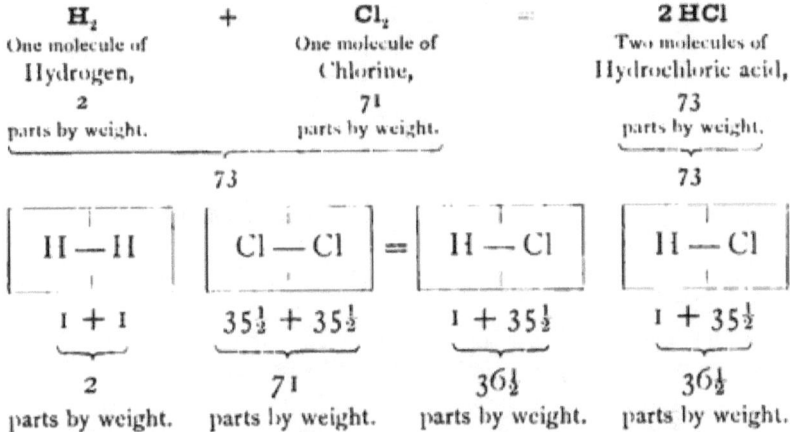

FIRST FACT. — Hydrogen gas and chlorine gas chemically combine.

SECOND FACT. — As a result they produce a new gas called hydrochloric acid gas, having properties different from those of the components.

THIRD FACT. — When hydrogen and chlorine combine, they do so in the proportions of two volumes of chlorine and two volumes of hydrogen.

FOURTH FACT. — The resulting gas has the bulk of four volumes; in other words, there is neither permanent contraction nor permanent expansion as a result of the act of union.

FIFTH FACT. — The combining volumes of hydrogen gas and chlorine gas have the weight-ratio of $2 : 71 = 1 : 35.5$. Now the accepted atomic weight of hydrogen is 1, and that of chlorine is 35.4.

Evidently then the facts stated sustain the statements of the law.

Second Illustration drawn from Water Vapor. — The following equations are applicable in this case:—

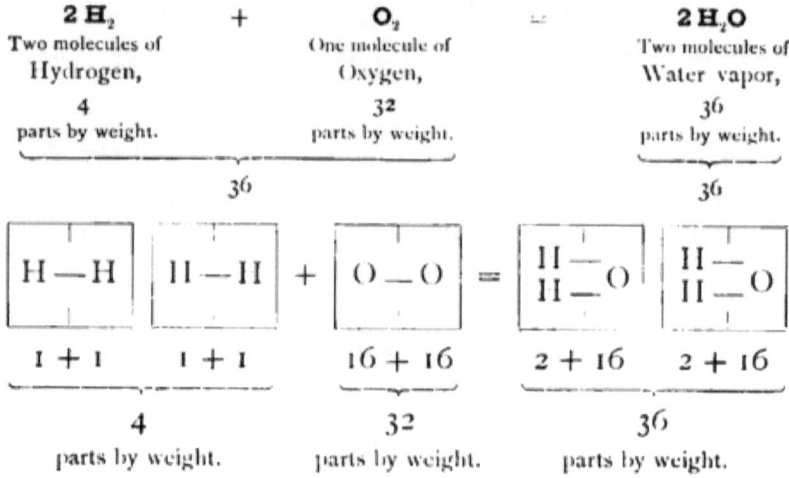

FIRST FACT. — Hydrogen gas and oxygen gas chemically unite.

SECOND FACT. — As a result they produce a new gas or vapor called hydric oxide or water vapor, having properties different from those of the component gases.

THIRD FACT. — When hydrogen and oxygen combine, they do so in the proportions of four volumes of hydrogen and two volumes of oxygen.

FOURTH FACT. — The resulting gas has the bulk of four volumes; that is, the same weight of matter has been packed into smaller space by the influence of the chemical union which has taken place. Thus there has been a permanent reduction from a total of six volumes to a total of four volumes. The ratio of $6:4 = 3:2$, and is a very simple one.

FIFTH FACT. — The combining volumes of hydrogen gas and oxygen gas have the weight-ratio of $4:32 = 2:16$. Now the accepted atomic weight of hydrogen is 1, and that of oxygen is 16.

Evidently, then, the facts stated under this illustration sustain the laws as given.

Avogadro's and Ampère's Hypothesis of the Size of Gaseous Molecules :—

Equal volumes of all substances, when in the gaseous state, and under like conditions of temperature and pressure, contain the same number of molecules.

Evidently the hypothesis declares that if, under certain conditions, one cubic foot of oxygen gas contains n molecules of oxygen, then under the same conditions one cubic foot of nitrogen gas, one cubic foot of hydrogen gas, one cubic foot of compound gases, as carbon dioxide (CO_2), ammonia gas (NH_3), each contain n molecules of their respective substances.

This hypothesis seems to follow directly from the laws of Boyle and of Charles. For if the material groups we call molecules exist at all, and if expansion and contraction of gases are in fact due to the moving apart or the moving together of these material groups, then the observed exact correspondence of the laws of such expansion and contraction, even in different substances, points conclusively to the existence of equal numbers of molecules in equal bulks of gases.

The hypothesis of Ampère is not the expression of a distinct and easily verified fact. It is rather the most reasonable explanation of a series of facts which cannot well be correlated without it.

As an example of its bearing, the following chemical illustration may be given: —

FIRST. Ammonia gas (NH_3) and hydrochloric acid gas (HCl) may be proved, by processes independent of this hypothesis, to have the formulas here assigned them.

SECOND. They are found by analysis to have the respective molecular weights 17 and 36.4.

THIRD. It is found experimentally that they always combine in the proportion of 17 parts by weight of ammonia gas to 36.4 parts by weight of hydrochloric acid gas. Hence they unite molecule for molecule.

FOURTH. It is found that they unite in equal volumes.

Whence these equal volumes appear to contain equal numbers of molecules.

The Relation of Diffusion of Gases to Ampère's Law.—That the facts relating to diffusion of gases afford a remarkable confirmation of the law of Avogadro and of Ampère, may be made apparent from the following example and explanations:—

Suppose a rubber balloon containing hydrogen gas and exposed to the air. The hydrogen gas is subject to two pressures from without,—the contractile force of the rubber, and the weight of the atmosphere. Why then is it not reduced to yet smaller bulk? Because of a resistance to pressure that it possesses in common with other forms of matter. In gases this resistance is called tension, or elasticity, or expansive power. But the facts of *diffusion* prove that all gaseous molecules are in particularly active motion, though with different rates. We are thence led to believe that the tension of the hydrogen in the balloon referred to is due to the impact of the moving hydrogen molecules; that is, to their outward blows against the enclosing walls of the balloon.

Again, suppose a balloon containing oxygen, but otherwise *in every way similar* to that containing hydrogen, just discussed. The portion of oxygen gas will have a weight sixteen times that of the hydrogen gas. The oxygen gas possesses tension, and this is due to the outward impact of the oxygen molecules against the walls of its containing vessel.

Now it is to be noted that in this second case the impact is equal to the impact of the hydrogen molecules in the other case; for in both cases the same external pressures are overcome.

The laws of mechanics show that the impact of any moving body may be expressed as equal to *one-half its mass multiplied by the square of its velocity.* Then the impact of a moving molecule of hydrogen and of oxygen may be expressed respectively as follows:—

i (impact of molecule of hydrogen) $= \frac{1}{2} mv^2$;
i' (impact of molecule of hydrogen) $= \frac{1}{2} m'v'^2$.

But it has been shown already that in the case of equal volumes of the two gases,

$$i = i';$$
whence $\qquad mv^2 = m'v'^2.$

Taking the velocity of the oxygen molecule (v') as the unit of comparison of velocity and calling it 1, and substituting for m and m' the ratios of their respective molecular weights, 1 and 16, we obtain,

$$v^2 = 16,$$
$$v = 4.$$

This result means that when the velocity of the oxygen molecule is called 1, the velocity of the hydrogen molecule is 4. Now this is in fact the rate of motion of the molecules as proved by Graham. And the result obtained from the course of reasoning here pursued has involved but one supposition; namely, that the two equal balloons, or in fact any two equal volumes of the gases, under like conditions of temperature and pressure, contain equal numbers of molecules.

The correct result attained contributes materially to place the hypothesis of Avogadro upon a mathematical foundation.

CHAPTER VIII.

CERTAIN FORMS OF ENERGY CLOSELY CONNECTED WITH CHEMICAL CHANGES.

HEAT AND ELECTRICITY.

It has been remarked that "a chemical operation presents two aspects to the investigator; it involves a change in the form or distribution of matter and a change in the form or distribution of energy."

Two forms of energy are especially involved in chemical changes: they are heat and electricity.

These subjects belong in a certain sense to the department of physics, yet by their sources, uses, and effects, they are so closely connected with chemistry that they admit of brief discussion here.

HEAT.

The invisible agency by whose transfer sensations of warmth and cold are produced, is itself called heat. Two kinds of heat may be distinguished:—

1. *Absorbed heat* is that which resides in a hot body, often remaining in it for a considerable time. It is transferred to another body, mainly by contact.

2. *Radiant heat* is heat in the act of passing with great velocity (about 190,000 miles per second) through space, whether vacuous or otherwise; radiant heat may

be either dark heat or luminous heat (the latter form being commonly known as light).

Theories of the Nature of Heat. — The dynamical theory of heat, and that now generally accepted, supposes that all matter as well as all space is pervaded by an extremely delicate and elastic medium called the *ether*. This theory regards (1) absorbed heat as a vibration of the molecules of matter; (2) radiant heat as an undulatory movement in the ether.

Heat as Motion. — That heat is not a form of matter appears to be shown by a variety of facts. For example, neither does addition of heat to a body increase its weight, nor does loss of heat by a body diminish that weight.

The quantity of heat in a given system is capable of indefinite increase; again, it can be destroyed as material substances cannot.

That heat is a form of energy — in other words, of motion — appears to be suggested by a multitude of phenomena. Of these a few may be mentioned.

FIRST. The general quantitative relations between heat and mass-motion are very simple. A given amount of mechanical motion may be changed into a certain definite amount of heat and no more; and on the other hand, a given amount of heat is capable of generating only a certain fixed amount of mass-motion.

Some of the contrivances ordinarily used to effect such interchange are imperfect and involve large losses during the transformations; these, however, are not losses of the total amount of energy, but only of that particular form of it which the appliance or machine may be intended to afford. Hence the strength of the argument is not impaired.

SECOND. The sources of heat are well explained by this view. They are friction, percussion, chemical action, the sun. (The internal heat of the earth need not be discussed here.)

Friction and percussion involve a diminution of mass-motion — or its entire quenching. But these have not been *destroyed;* they appear to have been merely transformed into minute molecular motions.

Chemical action evolves heat when certain substances combine. The true source of this heat appears to be that molecular percussion or atomic bombardment which is sustained when a myriad of atoms of one kind clash into combination with a myriad of another kind.

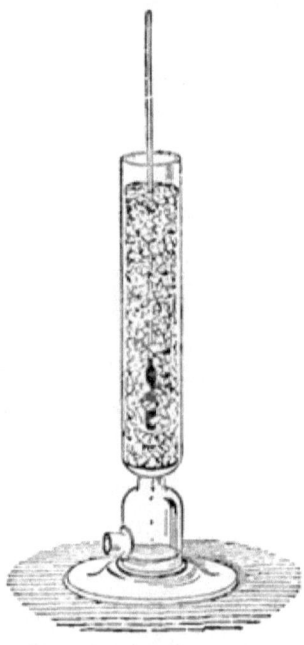

FIG. 47. — Apparatus for graduating a thermometer at the freezing-point of water.

The sun gives out an enormous amount of heat. Only a minute fractional part of it is received by this earth. (But this is a large amount as compared with man's ordinary means of producing energy.) But this tremendous and continual transfer does not appear to diminish the weight of the giver nor to increase that of the receiver.

Again, it appears more rational to believe that the incredible velocity of radiant heat is associated with a progressive flow of energy rather than with an actual transportation of matter.

THIRD. — The effects of heat are best explained by this view. The principal of these effects are the following: (*a*) temperature, (*b*) expansion and contraction, (*c*) change of state, (*d*) work, (*e*) light, (*f*) chemical combination and decomposition, (*g*) electricity.

(*a*) **Temperature.** — When a portion of matter gains or loses heat, the effect easiest and oftenest noticed is rise or fall of temperature. (But it is well known that in some cases this effect is altogether wanting. *See latent heat*, p. 47.)

What, then, is signified by the temperature of a body? Evidently its state of sensible thermal equilibrium or want of equilibrium as compared with some other body.

Temperature, then, is relative. It expresses the condition of a body in answer to the questions, does this body give heat to our persons? or, does it take heat from them? does this body give or withdraw heat from some other certain neighboring body with which it may be placed in contact?

To one of these questions our own nervous systems give a more or less distinct answer; to another we get an answer by observing some convenient secondary effect, such as the expansion of matter in some thermoscope.

The terms *warm* and *cold* mean, then, the transfer of a certain force in one direction or another toward a given body or from it. And it is supposed that this transfer results in an increase or a decrease of molecular motion.

But it is supposed that no known body exists in the condition of possessing absolutely no heat; that is, no molecular motion. Such a condition might be characterized as at a temperature of absolute zero (this temperature in assumed to be minus 273 degrees ordinary centigrade). Finally, it can easily be shown that temperature, whether judged by our sensations or by thermoscopes, is no certain index of the real amount of heat possessed by a body.

(*b*) **Expansion.** — With few exceptions all bodies, whether solid, liquid, or gaseous, expand with addition of heat and contract upon withdrawal of it. In these changes the heat certainly produces a mass-motion. It is observed in the alteration of the bulk of the whole mass of the body acted on, but apparently due to an increase or diminution of the motion of the molecules.

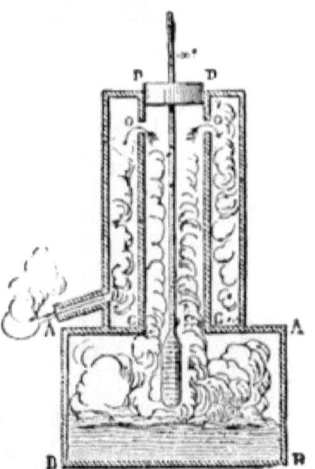

FIG. 48. — Steam box for use in graduating a thermometer at 100° C.

(*c*) **Change of State.** — When the molecular motion just referred to becomes so great as to carry the molecules of a solid or of a liquid beyond the range of those cohesive forces which characterize its previous state, it changes to the next less dense form of matter. In other words, addition of heat may in most cases change a solid to a liquid and a liquid to a gas. Again, withdrawal of heat may so restrict the molecular motion as to bring the molecules of a gas near enough together to make them subject to such cohesive forces as bind them into a liquid, or even to a solid.

There are some substances that have not yet been made to undergo change of state. Carbon, for example, has not yet been changed from the solid to the liquid form, — much less to the gaseous. But it can hardly be called an exception to the general statement. For in the first place the number of such substances is steadily decreasing; witness the recent liquefaction of the so-called permanent gases, oxygen, hydrogen, nitrogen, and the like. In the second place, all experience shows that whenever refractory solids are liquefied and vaporized, it is in consequence of addition of heat; and whenever liquids not previously frozen are solidified, it is in consequence of withdrawal of heat. (See p. 58.)

(*d*) **Chemical Combination and Decomposition.** — These subjects are important, but they are explained later.

(*e*) **Light.** — For the purposes of this book this subject may be presented in connection with spectrum analysis.

SPECTRUM ANALYSIS.

Spectrum analysis is much used in chemistry. It depends upon the following facts: —

1. Highly heated solids give out white light; *i.e.* a mixture of rays of light of various colors.

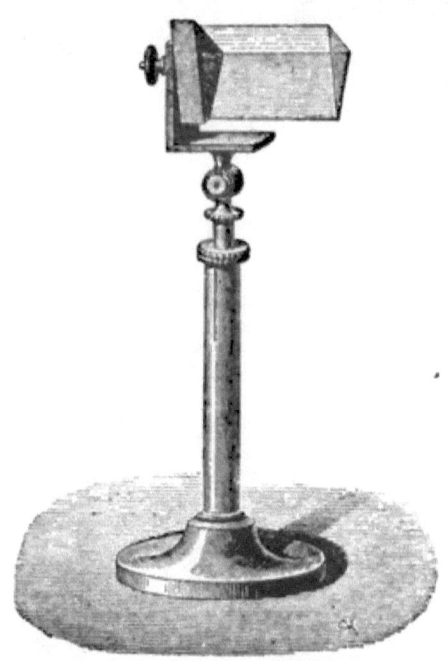

FIG. 49. — Glass prism in a convenient mounting for use in experiments upon the refraction of light.

2. Highly heated gases or vapors give out light of a characteristic color; *i.e.* either monochromatic light (composed of rays of a single quality), or polychromatic light (composed of rays of a character such that they cannot make up white light).

3. A ray of light when it passes through a prism is bent out of its course, *i.e.* suffers refraction; while a bundle of rays passing through a prism suffers dispersion; that is, the different rays of which the bundle is composed are bent out of course in different degrees.

The Spectrum. — The term *spectrum* is applied to that peculiar picture which appears when a narrow beam of light, after passing through a prism, is allowed to fall upon a screen.

Thus white light when passed through a prism gives a spectrum that contains an enormous number of rays. These rays are of all the different colors and shades of the rainbow, and they appear to be the same whatever the chemical constitution of the body giving out the white light. The spectrum is called a continuous one.

But colored light produces a discontinuous spectrum. Moreover, the

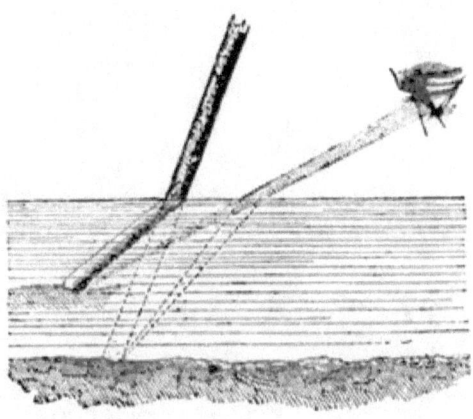

Fig. 50. — Refraction of light by water. A beam of light coming from the rod at the bottom of the water is bent out of course so that when it comes to the eye it appears to arrive from a different point, and thus the rod, in fact straight, appears to be bent.

character of such a spectrum varies with the chemical substance producing it.

The orange light emitted by highly heated sodium appears upon ordinary examination to be monochromatic. In a certain sense it is so. But critical investigation shows that its well-known color is made up of several minutely different shades.

The Spectroscope. — The following are a few of the fundamental principles upon which the use of the spectroscope depends: —

1. A spectroscope is a contrivance for examining light. In its simplest form it consists of three parts, — a narrow opening to admit the beam of light to be tested; a prism or series of prisms to disperse the rays of light; a small telescope to bring the spectrum to the eye.

2. A beam of light when passed through a suitable spectroscope yields a spectrum showing of what ray or rays the beam consists.

3. A beam of light may be tested just as it comes from its source; thus any substance may by some method be so heated as to become luminous, — that is, to give out light, — and such light may be passed through the spectroscope.

One method of heating is by an ordinary Bunsen lamp.

A second method applicable to solids and liquids is to cause some form

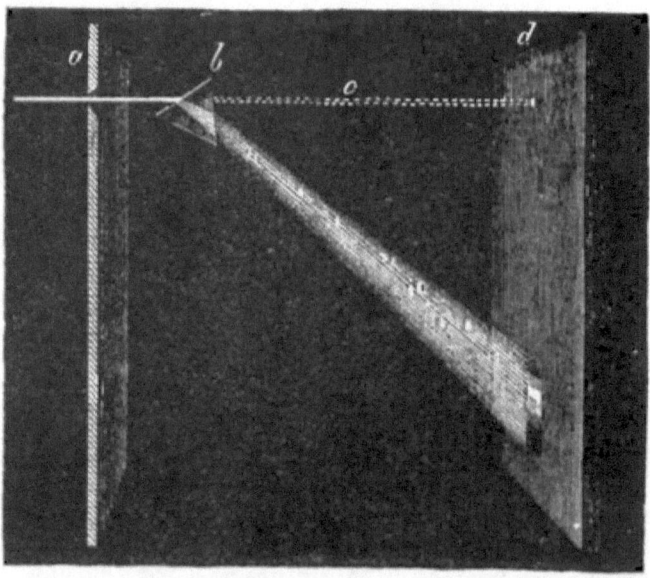

FIG. 51. — Refraction of light by a glass prism. A beam of light passing through an aperture at *a* would, if uninterrupted, fall upon the screen at *d*. If, however, it is interrupted by the prism *b*, it is bent out of course, *i.e.* refracted, and falls upon the screen at a lower point. At the same time the different rays composing the beam are refracted differently, thus producing a spectrum.

of electric discharge to flow from points of the solid to be tested, or over the surface of the liquid.

Gases may be made luminous by passing an electric discharge through glass tubes containing minute quantities of them.

4. Highly heated solids and liquids in general give out white light that is not characteristic, but is the same for all of them.

5. Highly heated vapors and gases in general give out light having colors that are peculiar, and characteristic of the atoms present.

Fig. 52. — Adjustable slit for use upon a spectroscope.

The denser the vapor, the greater the light; hence metallic vapors give out strongly luminous rays.

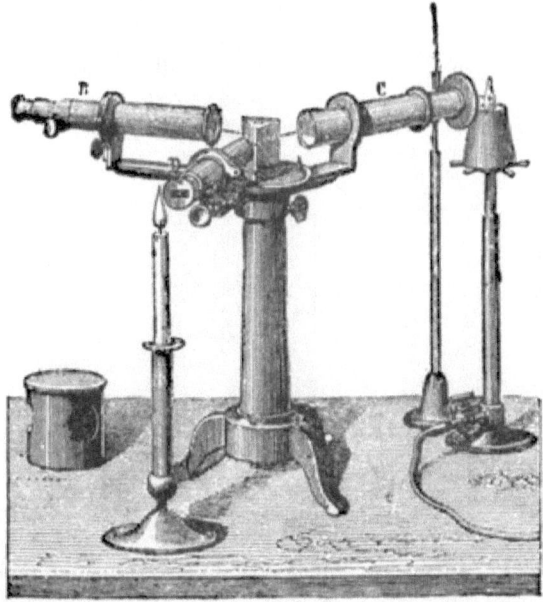

Fig. 53. — Spectroscope (of one prism) for the examination of luminous flames.

6. When light falls upon portions of matter, it is capable of at least three different dispositions: —

Certain substances allow nearly all the rays of a complex beam of light to pass through them.

Other bodies absorb certain rays and allow certain others to pass through them; the fact of such absorption is shown sometimes by the color of the body itself or sometimes by the spectroscope.

A third class contains bodies that are opaque; that is, they forbid any rays of light to pass through them, but they sometimes give back reflected light of a peculiar color.

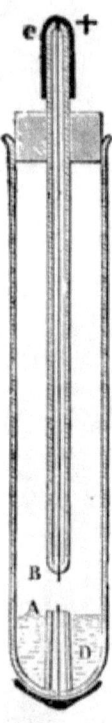

FIG. 54. — Apparatus for examining the spectrum of a liquid. Upon the passage of an electric current through the apparatus, the liquid is highly heated, and affords a luminous mass between the points AB. Thereupon this mass may be examined by the spectroscope.

The spectroscope is capable of showing whether light as originally emitted has been modified by the body upon which it has fallen.

As a result of the study of spectrum analysis it is believed that the light which characterizes each element is due to the atomic motion peculiar to that element. In fact, certain elements form colored compounds such as give similar spectral lines whether heated or cold, and so appear always to maintain the same rate of atomic motion (didymium).

That light is a form of motion is further sustained by the fact that spectrum lines appear displaced, owing to the rapid advance toward or retire-

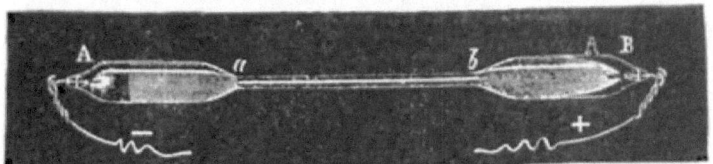

FIG. 55.—Geissler tube. It is so constructed that a portion of gas contained in it may be highly heated by an electric current. The narrow portion of the tube may be placed before the slit of the spectroscope for examination.

ment from the earth of certain comets and other celestial bodies. This displacement in the spectrum is analogous to the change of pitch of rapidly moving sonorous bodies, as the whistles on moving locomotives.

FIG. 56.—Direct vision spectroscope, employed to examine a colored liquid. The kind of dye-stuff in a liquid is often determined by this method of examination.

(*f*) **Work.**—Heat may accomplish external or internal work. The former product is evident to us in various visible forms of mass-motion.

Internal work is done when some natural molecular force is overcome. Thus this is accomplished when the cohesive power of a solid is so overpowered that liquid is produced. Ice when melted occupies a diminished bulk, so that in this case no external work results.

FIG. 57.—Spectroscope attached to a telescope for the purpose of examining the protuberances on the outside of the disk of the sun. (The portrait represents J. Norman Lockyer, the celebrated English astronomer.)

The internal work of heat is explained by supposing that in this case there is produced some internal motion, like rotation of molecules, for example, rather than one like a translation of them.

ELECTRICITY.

The phenomena of electricity already known are so numerous, varied, and subtile as to defy complete explanation. Of the various theories of its nature that have been suggested, none seem on the whole so satisfactory as that it is some form of energy; in other words, a form of motion. The exact character of this motion cannot at present be stated. It is probable that in some cases it is a motion of translation of molecules; oftener, perhaps, it is a motion of mere rotation or similar polarization of molecules without change of position. Many of the phenomena seem to demand the intervention of some medium like the ether, — whether the same as that supposed for the explanation of radiant heat or only a somewhat similar one, it is impossible to declare. However objectionable the theory of an ether — or of one or more fluids — is, yet in the present state of knowledge something of the sort seems indispensable for purposes of explanation.

Note the tendency to use such expressions as "the electric current." That there is any current of matter is not probable; but there certainly is a progressive transfer of electric energy. This transfer is marked by a velocity nearly double that of light.

Its Sources. — It is a very suggestive fact that whatever known source of electricity is invoked, there is always an evident consumption of force. This becomes apparent from a mere enumeration of some of its direct sources.

Such sources are: the energy of friction; the energy of mere mechanical separation of certain bodies in contact; mere application of heat;

FIG. 58. — View illustrating the use of a waterfall for actuating dynamos to produce electric currents for the electric lighting of a city.

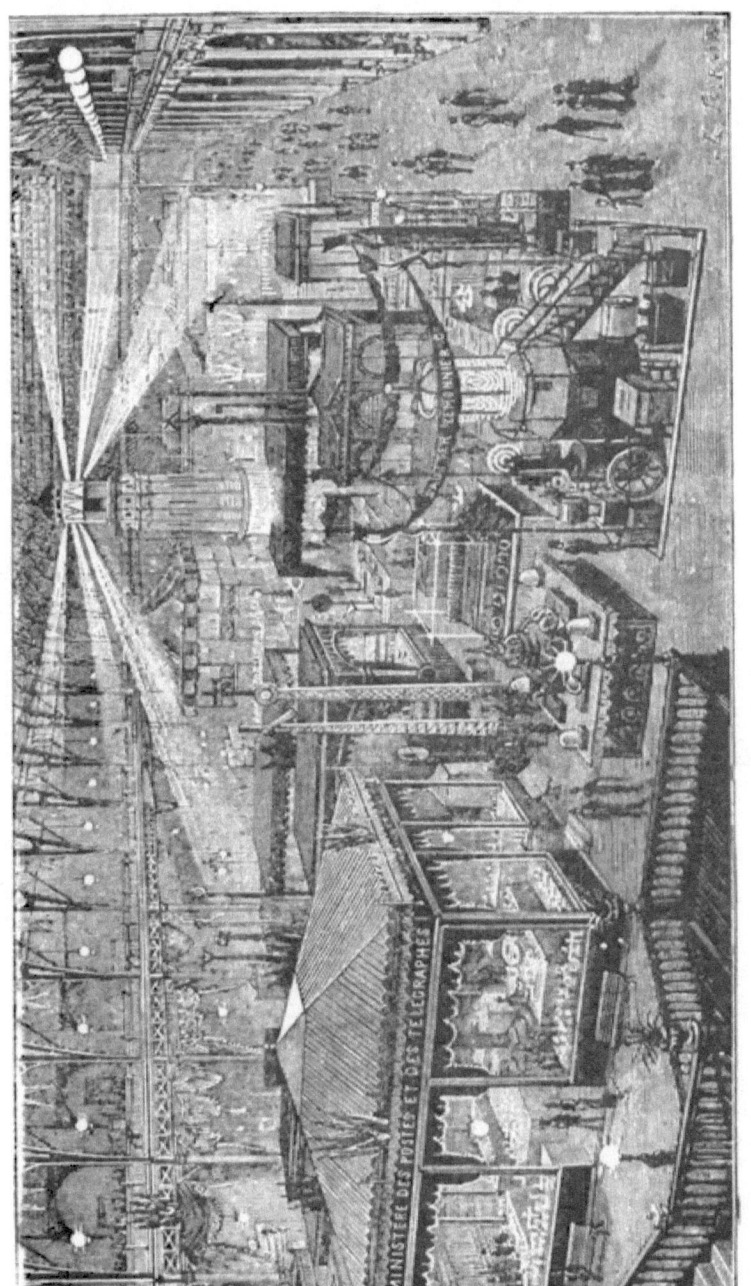

Fig. 59. — View in an exhibition of various forms of apparatus employing the electric current for useful purposes.

mere change of temperature; mechanical separation of magnetically attracted bodies; chemical combination and decomposition.

The modern system of producing electricity for purposes of illumination illustrates, in an interesting manner, several transformations of force.

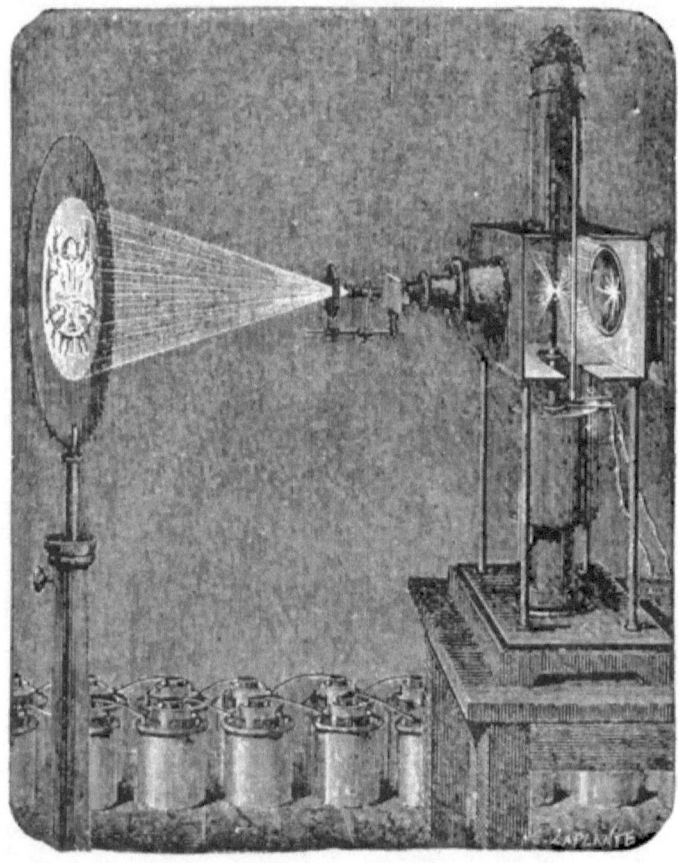

FIG. 60. — Electric light produced by a Bunsen battery and employed in a magic lantern for the purpose of projecting the image of a microscopic object upon the screen.

This system generally includes at least four contrivances: first, the fire-box; second, the boiler and engine; third, the dynamo; fourth, the lamp. In the fire-box, energy of the chemical affinity of burning coal is transformed into the energy of heat; in the boiler and engine the energy of heat is transformed into the energy of mass-motion; in the dynamo the energy of mass-motion is transformed into the energy of electricity; in the lamp, the energy of electricity is transformed into the energy of light and heat.

Its Effects. — That the effects of electricity are strikingly suggestive of some form of motion is evinced by a brief reference to a few of them.

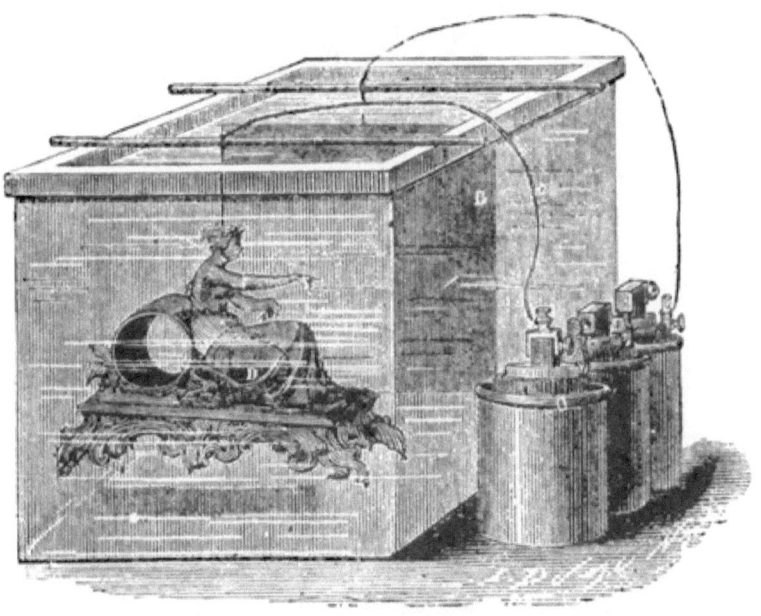

FIG. 61. — Method of precipitating silver or gold from a solution of the metal upon an ornamental object. In this case the electric current from a Bunsen battery is employed. The current from a *dynamo* will also serve.

(*a*) One of the most early and easily observed of the effects is mechanical motion; thus electrified bodies readily exhibit direct attraction and repulsion of mass. Mechanical apparatus can also be kept in motion by a proper application of electricity.

(*b*) Electricity gives rise, by direct transformation, to other admitted forms of force, such as heat, light, and magnetic polarity.

(*c*) Electricity produces motion in chemical molecules. In the processes of electro-plating and other forms of electrolysis it effects a drawing apart of portions of matter previously firmly bound together by chemical affinity.

(*d*) The very peculiar phenomena of conduction, insulation, and electric induction, seem to favor the view here presented; for it is hardly possible to conceive of such results springing from a transfer of matter, even of a most tenuous kind.

CHAPTER IX.

THE ATTRACTIONS OF MASSES.

GRAVITATION.

GRAVITATION is a force by which considerable portions of matter are mutually attracted, whatever their size, nature, distance apart, or the intervening medium.

Law.—*The gravitating attraction between material objects is directly proportional to their masses, and inversely proportional to the squares of the distances between their centres of gravity.*

In accordance with this law, then, if two bodies, each containing a unit of mass, gravitate toward each other with a certain force, then when one of those bodies has its mass doubled or trebled, the gravitating force is multiplied by two or three; and if at once the mass of one body is doubled, and the mass of the other is trebled, then the gravitating force is doubled by reason of the one increase, and trebled by reason of the other; that is, it is increased in the proportion of $2 \times 3 = 6$ times.

Again, if the masses remain unchanged, and the distance between the *centres of gravity* of the bodies is increased to two or to three units of distance, then the gravitating force is reduced respectively to

$$\frac{1}{2^2} = \frac{1}{4}, \text{ or to } \frac{1}{3^2} = \frac{1}{9}.$$

If the distance stated is reduced to one-half or one-third, then the gravitating force is increased to four times or nine times, respectively.

Gravitation is the great agent of stability in nature. It regulates the motions of the heavenly bodies. By its influence objects on the surface of the earth retain their positions instead of being cast forth into space.

Its value is best estimated by considering the results which would follow its suspension, supposing the latter to be possible — though in fact it is not.

CHAPTER X.

THE ATTRACTIONS OF MOLECULES.

I.—COHESION.

COHESION is an attractive force acting at insensible distances between molecules of the *same kind*. Besides

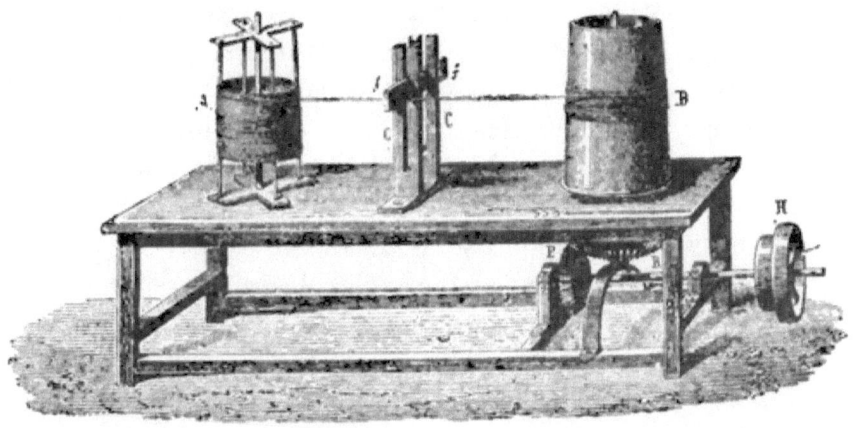

FIG. 62.—Wire-drawing machine. The coarser wire on the reel *A* passes through a small opening in the steel plate *f*. It is wound as a finer wire upon the drum *B*. The operation illustrates a resistance of solids to change of form. (The metal remains solid during the operation.)

this cohesive force, molecules of the same kind are influenced by an expansive tendency (undoubtedly due to heat).

In *solids* the cohesive force manifests itself in the resistance they offer to any derangement of the shape of the solid; that is, of the arrangement of molecules within the mass. Thus solids offer resistance to increase or decrease of volume, also to flexure and to rupture, and, indeed, to any alteration of shape, even though not involving increase or decrease of bulk.

In *gases* the opposite extreme is found. They seem to possess little or no cohesive force; the expansive tendency predominates. A few cubic inches of gas placed in a vacuous receiver, of any shape, and of any size (not exceeding forty or fifty miles in height), would doubtless soon expand so as to fill the receiver. Apparently the remoteness of the molecules of a gas from one another is an important feature; they are thereby removed more from the influence of their specific cohesive forces, and thus are made capable of greater freedom of motion.

Liquids exist under conditions that are in a certain sense intermediate between those of solids and of gases.

Liquids manifest cohesion in their tendency to assume the globular form (that form affording the most compact arrangement for a given number of centrally attracted objects).

Liquids display the existence of the expansive tendency in their easy evaporation.

Polarity. — There is a striking difference between solids and liquids as to the adjustments of the molecular forces. In liquids the cohesive forces seem to be balanced about the centres of the molecules, so that these molecules are free to move about each other, and to occupy equally well many positions with respect to each other. In solids, on the other hand, there exists *polarity; i.e.* the cohesive forces seem to reside out of the centres of the molecules, and in certain centres of force which may be called poles.

This polarity not only compels solids to offer that resistance to derangement of shape already referred to. It also produces the phenomena of crystallization.

Crystallization. — In general, a portion of matter in the act of changing from the gaseous or the liquid to the solid form manifests a tendency toward a definite arrangement of molecules. With the exception of a few animal and vegetable products every solid affects a definite polyhedral form, although it may manifest this tendency only under favorable circumstances. A body possessing such a form is called a crystal.

It seems proper to assume that the crystalline condition is the normal one for all solids.

It is true that certain distinctively animal and vegetable matters — the so-called organized matters — assume the cellular form rather than the crystalline. They are not as thoroughly exceptional, however, as might at first appear. (See p. 105.)

Cleavage. — Certain well-crystallized substances, when broken, are found to possess the property of cleavage in a marked degree: if pulverized, or otherwise subdivided, they undergo fracture with far greater ease in certain directions than in others. A given portion of Iceland spar, for example, having a well-defined rhombohedral form, may be broken up into smaller rhombohedrons rather than into masses of indefinite shape. So a mass of galena, if broken, forms rectangular or cubical masses. Even when these substances are pulverized, examination by the microscope shows, in each case, the strongly marked crystalline tendencies; each particle of powder shows itself to be a little crystal, or mass of crystals. The same principle is observed in other crystalline substances, though perhaps in less marked degree.

In general, it may be assumed that any crystalline substances when broken up into small particles will show crystalline fracture, and in the powder will show a multitude of little crystals bearing a strong resemblance to the large crystal whence it came. The cutting of diamonds and other precious stones depends upon this general principle; *i.e.* that layers may be cleaved off in certain directions better than in others. It is very plain, then, that a crystal possesses not merely external symmetry — the laws of its structure govern most thoroughly its internal parts. It may be imagined that, if it were practicable to reduce the size of a given large crystal by removal of its outer layers, one by one, the time would come when, the limit of the single molecule being reached, this molecule itself — if capable of remaining solid — would be found to possess something equivalent to a crystalline form. It would be either a miniature of the large crystal or else one geometrically related to it.

Theoretically, at least, it may be considered that a single molecule first solidifies, and then other molecules build themselves upon it as upon a nucleus. It may be assumed that they pile themselves up upon these outer faces in accordance with some definite geometrical law. These statements strengthen the conviction that crystalline form is not a mere external and superficial peculiarity of substances, but is the product of a fundamental and essential law of them. Probably it is closely connected with their atomic and molecular constitution. Mitscherlich's law (see p. 234) represents an attempt to deal with this relationship. It may be expected that in the future yet other and more definite connection will be shown between the crystalline form and chemical constitution.

Apparent Exception. — It is commonly accepted as a principle that in a certain sense organized bodies (see p. 154) do not conform to these statements. Perhaps in another sense they will ultimately be found in harmony with it. Thus certain organized bodies, like muscle, are made up, not of one substance, but of several different substances, arranged in the cellular form. If these substances could be separated one from another, and then each reduced to the solid form, perhaps they would then appear as a set of crystalline compounds. This latter statement is made with the general admission, already presented in another place, that different substances well recognized as crystallizable assume the crystalline form with different degrees of ease. In other words, in order to crystallize, substances demand conformity to many conditions.

Crystalline Systems. — The various crystalline forms recognized have been classified in six so-called *systems;* and, moreover, a substance crystallizing in a given system is capable of certain variations within that system. In this way there exist within a given system not only the simple forms, but also what are called *compound forms* and *hemihedral forms.*

Certain substances crystallize in shapes that are readily recognized and referred to their proper systems. Others assume such complicated forms, or else such imperfectly developed shapes, or else crystallize in such minute portions, that their recognition is difficult. Sometimes a special apparatus called a goniometer is employed to determine the particular crystalline form.

The several systems are presented in brief but compact form in the following tables and diagrams : —

FIRST SYSTEM.
REGULAR OR ISOMETRIC.

Three rectangular axes, all equal.

ALLIED FORMS.

Cube. Regular octahedron. Tetrahedron. Rhombic dodecahedron. Trapezohedron.

EXAMPLES.

Diamond; alum; fluor spar; common salt; iron pyrites; garnet; copper; nickel; gold; lead; arsenious oxide.

FIG. 63.

SECOND SYSTEM.
QUADRATIC OR TETRAGONAL.

Three rectangular axes, two equal.

ALLIED FORMS.

Square prism. Octahedron with a square base.

EXAMPLES.

Tin; potassic ferrocyanide; mercuric cyanide; mercurous chloride.

FIG. 64.

THIRD SYSTEM.
HEXAGONAL.

Three equal axes, not rectangular, intersecting at 60°. One perpendicular to the three.

ALLIED FORMS.

Rhombohedron. Bipyramidal dodecahedron. Hexahedral prism.

EXAMPLES.

Graphite; ice; quartz; amethyst; corundum; emerald; calc spar; tourmaline; camphor; sodic nitrate.

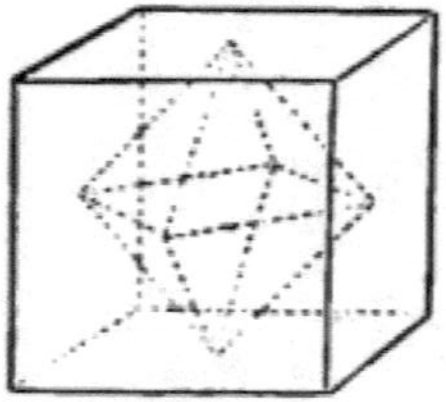

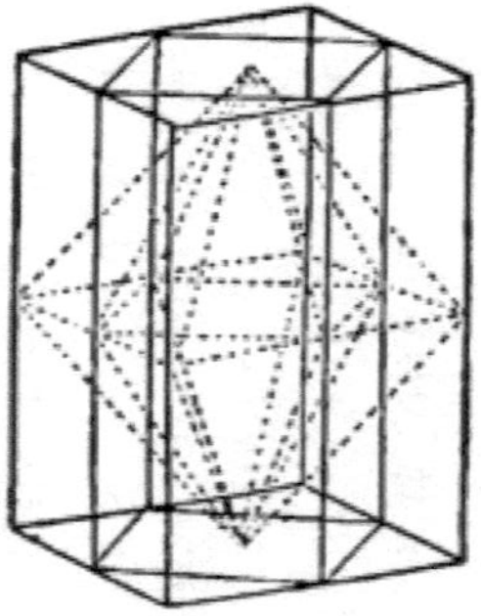

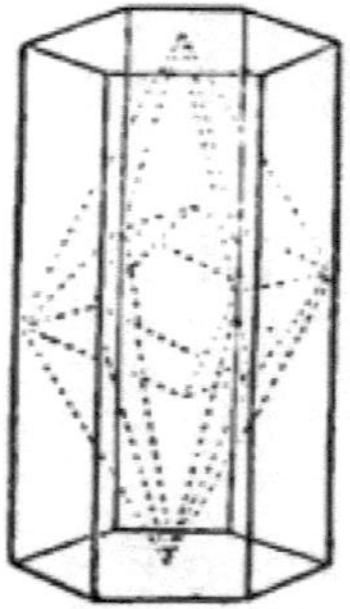

FIG. 65.

THE ATTRACTIONS OF MOLECULES.

FOURTH SYSTEM.
RHOMBIC OR ORTHORHOMBIC.

Three rectangular axes, no two equal.

ALLIED FORMS.

Right rectangular prism. Right rhombic prism. Right rectangular octahedron. Right rhombic octahedron.

EXAMPLES.

Sulphur; topaz; aragonite; baric sulphate; potassic sulphate; magnesic sulphate; zinc sulphate; potassic nitrate; potassic dichromate; rochelle salt; citric acid.

Fig. 66.

FIFTH SYSTEM.
MONOCLINIC OR MONOSYMMETRIC.

Two axes oblique, the third perpendicular to both.

ALLIED FORMS.

Oblique rectangular prism. Oblique rhombic prism. Oblique rectangular octahedron. Oblique rhombic octahedron.

EXAMPLES.

Sulphur; felspar; mica; epidote; calcic sulphate; sodic sulphate; sodic carbonate; potassic chlorate; borax; sugar; oxalic acid.

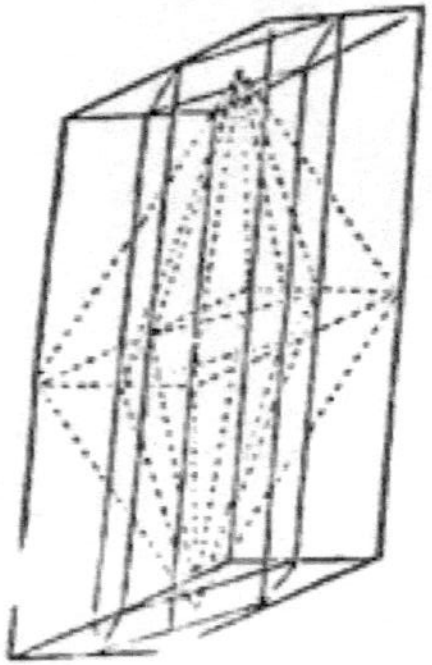

Fig. 67.

SIXTH SYSTEM.
TRICLINIC OR ASYMMETRIC.

Three axes, all intersecting each other obliquely.

ALLIED FORMS.

Doubly oblique prism. Doubly oblique octahedron.

EXAMPLES.

Ferrous sulphate; cupric sulphate; bismuthyl nitrate; sulphate of cinchonine.

Fig. 68.

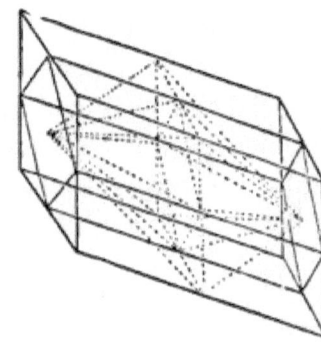

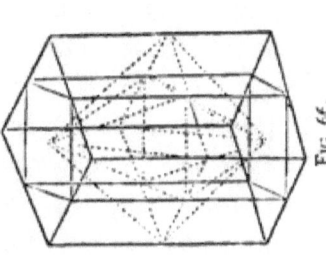

The Process of Crystallization. — The process of crystallization is a variety of the process of solidification; but it is one that is dependent upon a highly specialized *arrangement* of the particles of the solid. Hence it must be supposed that solidifying molecules take up certain determinate motions before they have placed themselves in those symmetrical and orderly

Fig. 69. — Crystals formed in a mass of metallic bismuth by slow cooling of the melted metal.

ranks required by the crystalline condition. This special kind of motion is not consistent, however, with the two fluid states of matter, nor yet with the rigid one. It seems to best take place in that transition period which extends between them. It might be expected, then, that crystallization would be facilitated by increasing as far as possible the extent of this transition state. This is found to be, in fact, the case. Whenever the progress of solidification is prolonged, the tendencies toward crystallization are favored.

THE ATTRACTIONS OF MOLECULES. 109

Crystals are usually produced by the following means: —

(*a*) **The Slow Cooling of Vapors.** — The formation of crystals of snow from water-vapor in the atmosphere is an example of this method.

The formation of crystals of iodine is another example.

FIG. 70. — Section of a crucible in which melted sulphur has been allowed to crystallize by slow cooling.

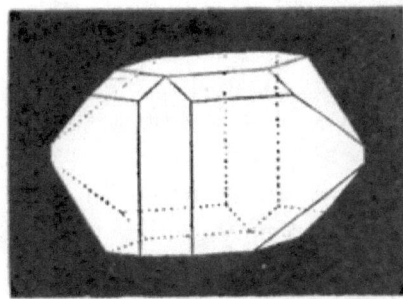

FIG. 71. — Diagram showing crystalline form assumed by sulphur during slow cooling from the melted form.

(*b*) **The Slow Cooling of Liquids produced by Fusion.** — The crystallization of melted sulphur, of melted bismuth, and of melted zinc are examples of this method.

(*c*) **The Slow Cooling of Liquids produced by Solution.** — This is the method by which most crystals produced in the arts are formed. A very familiar example is rock candy, which is crystallized by cooling the liquid produced by dissolving cane sugar in hot water.

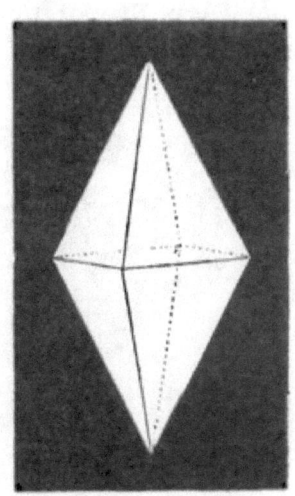

FIG. 72. — A diagram showing crystalline form in which sulphur is found in nature.

Multitudes of crystalline salts are manufactured by chemists by this method.

Fig. 73. — Method of producing crystals of rock candy by slow cooling of the solution of cane sugar in water. The crystals collect upon threads stretched through the liquid.

(*d*) **Slow Evaporation of Liquid Solutions.** — The majority of solid chemical salts known may be dissolved in

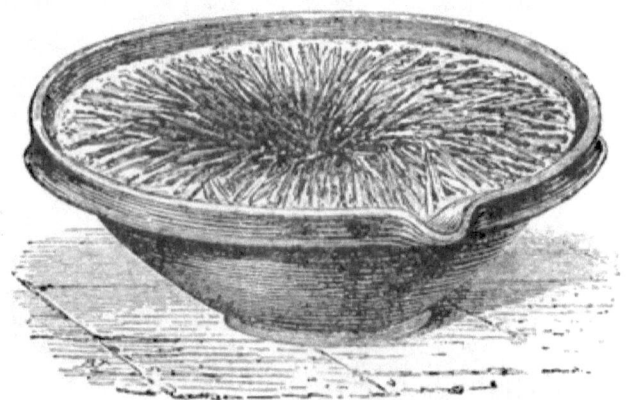

Fig. 74. — Crystals of potassic nitrate (saltpetre) formed by the slow cooling of a solution of the salt in boiling water.

water and thus changed temporarily to the liquid form. If such a liquid is evaporated, the water may be expelled

FIG. 75.— Artificial reservoirs constructed on the margin of the Mediterranean for the purpose of evaporating sea-water in order to obtain salt. When the water is sufficiently concentrated by the heat of the sun, the salt crystallizes, and is raked out upon the dikes separating the reservoirs.

and the solid caused to reappear. If the expulsion of the water is conducted very slowly, the solid reappears so slowly that its molecules have time to arrange themselves in the form of crystals.

An example of this method is found in the manufacture of common salt, which is generally crystallized from its solution in water, by means of slow evaporation.

FIG. 76. — Rock crystals found in nature. The substance is silicic oxide (SiO_2). It forms in hexagonal prisms surmounted by hexagonal pyramids.

Theoretically, a crystal once formed in a solution and continuing to increase in size (by every face of every set of faces receiving a deposit of the same thickness) would produce an ideally perfect crystal. But as a fact, distortion almost always results. By the change in specific gravity in the liquid, from loss of the solid substance, currents are generated, and different parts of the crystal are subjected to different conditions. Again, by the crystal becoming attached to other crystals, or to the side or bottom of the vessel, the several faces receive unequal deposits; yet every face always remains parallel to its original position and the interfacial angles are constant.

CHAPTER XI.

THE ATTRACTIONS OF MOLECULES (*continued*).

II.—ADHESION.

ADHESION is a form of attractive force, exerted at insensible distances, between molecules of different kinds.

(*A*) ADHESION BETWEEN SOLIDS AND SOLIDS.

Of this kind of adhesion there are many examples. The adhesion of a piece of wood to another of different kind by means of a layer of glue involves two illustrations; for the solidified glue adheres to each kind of wood.

The rock known as granite is composed of three different and separate materials, — quartz, felspar, and mica, — easily recognized by inspection; they are held together by a form of adhesion.

There are certain cases in which two solids when brought in contact liquefy, or set up some very evident chemical change. Thus when solid ice and solid salt are placed together, each exerts on the other a very remarkable attractive force by reason of which the ice melts and the salt dissolves in the water formed.

The phenomena of this operation, and others similar to it, are discussed more properly under Dissolving of Solids in Liquids (p. 118) and Chemical Action (pp. 169 and 171).

(*B*) ADHESION BETWEEN SOLIDS AND LIQUIDS.

Of this kind of adhesion there are many forms worthy of consideration here.

First Form. — *Moistening.*

This form is exemplified by many solids and liquids. Thus, a glass rod dipped in water and then withdrawn is found to retain some water upon its surface.

FIG. 77. — Disposition of apparatus for showing the adhesion of a solid to the surface of a liquid. The upper part of the figure represents one pan of the balance. In the other pan of the balance (not shown in the cut) weights may be added until the plate (shown at the bottom of the cut) is pulled away from contact with the liquid.

Second Form. — *Capillary Attraction.*

When a tube, open at both ends, is dipped into liquid contained in a considerably larger vessel, there may be three cases (under this title).

116   THE ATTRACTIONS OF MOLECULES.

The case oftenest observed is that in which the liquid in the tube rises to some distance above the general level of the liquid in the vessel. A tube of glass dipped in water displays this phenomenon.

In some cases the liquid in the tube is depressed below the level of that in the vessel. A glass tube dipped in mercury affords an illustration of this case.

It might be expected that the case of a liquid maintaining the same level within and without the tube

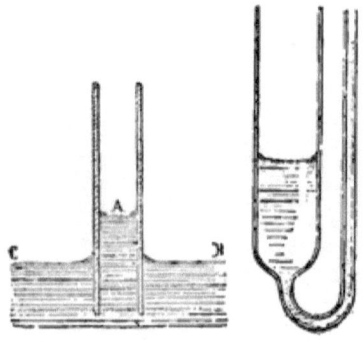 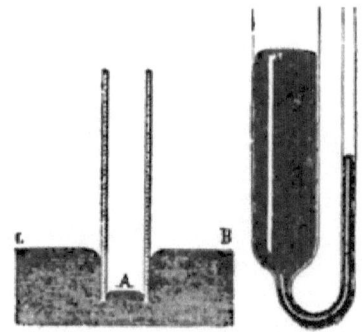

FIG. 78.—Capillary attraction, showing rise of liquid in narrow tubes.   FIG. 79.—Capillary depression, shown by fall of mercury in a narrow tube of glass.

would be a rare one; for this can only exist when there prevails a certain exact balance between the amount of cohesion of the liquid itself, and double the amount of the adhesive force of the liquid to the material of the tube. Of course exact equality is everywhere an exceptional condition of things.

THIRD FORM. — *Spheroidal State.*

When a small portion of liquid is placed upon the surface of a supporting material that is relatively highly heated, the liquid draws itself up into a globule and moves about in what is called the spheroidal state. The heated surface causes the liquid to evaporate chiefly on its under side; the abundant vapors thus

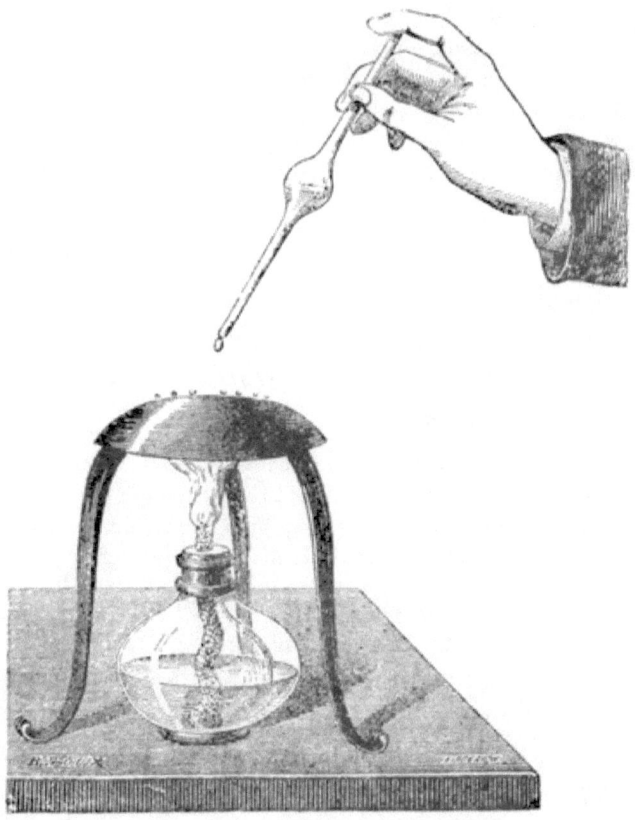

Fig. 80. — Experiment to illustrate the spheroidal state of water. The lamp heats the plate, whereupon, if drops of water are placed upon it, they remain as little spheres and do not adhere to the plate.

produced afford a cushion upon which the liquid is supported. Of course the liquid does not rest in actual contact with the heated material.

A noticeable anomaly exists in the case described. Even upon a solid surface of a very high temperature the liquid does not evaporate as rapidly as in a vessel sustained at a much lower temperature. But at these lower temperatures the liquid rests *in contact* with the solid, and the entire mass of liquid receives heat by the processes of *conduction* and *convection*. At very high temperatures, however, the liquid is not in contact; the globule then receives heat just as other objects do at a great distance from a source of heat; that is, by the process called *radiation*. By this means, the lower surface of the globule is the portion chiefly influenced; here vapors are given off in abundance — sufficient to afford the supporting cushion, but not sufficient to rapidly diminish the mass of the globule.

FOURTH FORM. — *Solution of a Solid in a Liquid.*

The dissolving of one or more liquids may involve the interaction of a great many forces, so that solution in its more complex varieties is worthy of careful study and extended discussion.

The general opinion now prevails that substances in dilute solutions exist in a condition somewhat analogous to substances in the gaseous state under moderate pressure and moderately high temperature. This view is based upon the studies of J. H. van 't Hoff and F. Raoult.

Even in the simplest forms the process of solution depends on a variety of conditions.

The *amount* of a given solid capable of dissolving in a liquid, in a given experiment, depends upon at least three factors : —

The nature of the solid and the liquid ;

The quantity of the solid and the liquid ;

The temperature under which the experiment is conducted.

The *rapidity* with which in a given experiment any certain (possible) amount of a solid is dissolved in a given amount of liquid, depends upon the *rapidity* with which the favorable conditions are provided. Thus extreme fineness of comminution, agitation of the mixture of solid and liquid, rapid addition of heat, — all favor rapid dissolving.

1. **Nature of the Liquid and the Solid.** — (*a*) Water is specifically gifted with solvent power of a remarkably wide range. Moreover, it is a very abundant and widely diffused substance. In many ways it seems entitled to be considered the chief liquid. Now water dissolves in large quantities a very large number of chemical salts. As examples may be mentioned, potassic sulphate, K_2SO_4; sodic sulphate, $Na_2SO_4 \cdot 10\,H_2O$; potassic nitrate, KNO_3; zinc sulphate, $ZnSO_4 \cdot 7\,H_2O$; sodic chloride, $NaCl$. In fact, it dissolves in greater or smaller quantity the majority of salts known. Water also dissolves a great number of neutral bodies, of which cane sugar ($C_{12}H_{22}O_{11}$) may be used as an example.

(*b*) Carbon disulphide dissolves sulphur and some other substances not soluble in water.

(*c*) Liquid oils dissolve many solid fats; thus the more liquid paraffins dissolve the more solid ones like white paraffin wax.

(*d*) Liquid mercury dissolves most of the metals, as potassium, sodium, gold, silver, zinc (but not iron). Mercurial solutions and their more solid forms are called amalgams.

(*e*) Melted zinc dissolves many metals, as solid copper, gold, platinum, and others. Products of this general character both before and after solidification are called alloys.

(*f*) Diluted sulphuric acid dissolves metallic zinc and other metals.

Remarks on this List. — This list is merely one of examples. Other examples might have been given under each head, and perhaps with equal propriety.

In examples *b, c, d, e*, it plainly appears *that solvent power is generally the greater, the greater the similarity of the solid and the solvent liquid.*

It will be noted that, as a matter of course, solvent action is oftenest observed in cases of liquids which (like water, for example) remain in the liquid form at the temperatures ordinarily prevailing; it must not be forgotten that if the globe had a slightly lower climatic temperature, many of these would be best known as solids; they would be thus reduced to the category of those (like metallic zinc, for example) which now have to be artificially raised a little in temperature before they display their solvent powers. Of course, at greatly reduced temperatures, all known liquids solidify, and would be thrown out of the account, just as at very greatly elevated temperatures all known solids would probably liquefy, and so would come into the list of liquid solvents.

Example *f* needs special attention. In a very just sense it belongs to case *a*. The solvent action is properly that of *water upon a chemical salt* — zinc sulphate ($ZnSO_4 \cdot 7H_2O$). For it happens that the dilute sulphuric acid exerts such a chemical action upon the metallic zinc as changes it into the chemical salt — zinc sulphate. When dilute sulphuric acid acts upon metallic zinc, the operation may be represented as follows: —

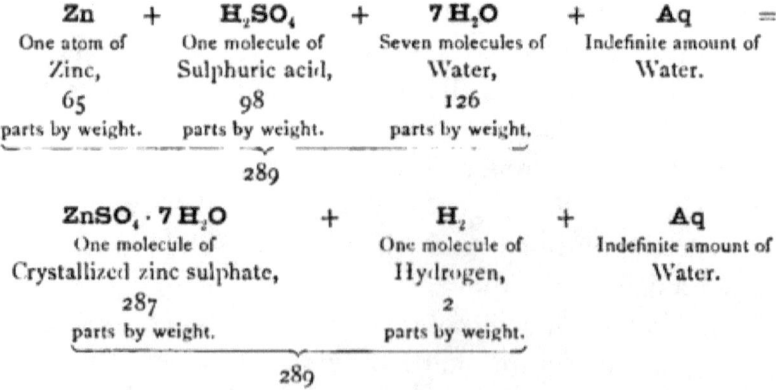

Then the zinc sulphate ($ZnSO_4 \cdot 7H_2O$) dissolves in the water present in the original dilute sulphuric acid.

It ought to be noted that it seems highly probable that in all cases of solution — even those where a single solid dissolves in a single liquid —

there is some chemical action. In cases at one extreme of the series the action may be very marked. This is so, for example, when sulphuric oxide (SO_3), a solid, dissolves in water; great heat is here evolved, and there is produced a new substance, sulphuric acid (H_2SO_4), which also dissolves in water. An example of the opposite extreme is that in which sugar dissolves in water. It would be difficult in such a case to demonstrate that any chemical action takes place unless specially devised means were employed for its detection.

2. **Influence of Change of Temperature.** — The comprehension of this, as well as of other branches of the subject, may be facilitated by a description of the operation of solution as advancing by stages.

When a solid is placed in a liquid, the liquid acts first upon the outer layers of the solid.

Adhesion and chemical action are exerted at once. The surface molecules of the solid leave the others and assume the liquid form.

The liquefied portions from the original solid at once diffuse themselves into some of the intermolecular spaces of the bathing liquid.

True solution has now been accomplished, though it may as yet be somewhat limited quantitatively.

By the very act of dissolving, as thus far described, there have been created certain conditions which directly oppose its further progress.

The first of these opposing conditions is a reduced temperature. It is a recognized law that in all changes of a solid to a liquid, absorption of heat takes place. This effect is recognized in the cooling of whatever happens to be the surrounding medium. It is often stated that in liquefaction sensible heat becomes *latent heat*. (While the expression *latent heat of liquefaction* is a well-established one, it is somewhat inappropriate. It carries the suggestion that a certain amount of heat has become merely *concealed*, whereas heat in fact *ceases to be* — as heat — when it does the work of liquefying a solid.) To the foregoing should be added another important statement. It is the following: The amount of a solid that a liquid can hold in solution varies with the temperature, being in most cases the greater the higher the temperature. It follows that at any given point of temperature the liquid may hold dissolved a certain amount of the solid and no more. When this full amount is in fact dissolved, the liquid is said to be *saturated*. In case of many solids and liquids, tables have been constructed showing by the graphical method the quantities producing saturation at each of a series of temperatures.

From what has been said it is plain that if it is desired that the dissolving operation shall go on, heat must be added.

The second of the opposing conditions is the local saturation of the solvent liquid. By reason of this saturation the portions of liquid lying immediately about the solid may become incapable of further dissolving action, while more remote portions may not as yet have begun to act. This cessation is sometimes associated with the fact that the solid rests at the bottom of the liquid, and the solution, being heavier than the original liquid, naturally rests upon and covers the solid. There are three ways of overcoming this difficulty. One way is to agitate the liquid mechanically. A better way is to suspend the solid in a perforated basket hung in the upper layers of the liquid; then the solution, as it becomes saturated, sinks by its own weight, and is promptly replaced by portions of fresh liquid. A third method, employed to advantage in combination with the second, is heating; this produces convective currents in the liquid.

3. **The Quantity of the Solid and Liquid.** — This point needs no discussion in addition to the statements relatting to *saturation* already introduced.

Deliquescence. — This is a form of solution. It is exemplified by certain substances that have such a strong attraction for water that they even absorb that moisture existing as vapor in the atmosphere. They draw this water to themselves in such quantity that they soon cease to be solid; for they liquefy by dissolving in the water absorbed.

This topic is naturally associated with the adhesion of *solids and gases.* (See p. 130.)

Freezing Mixtures. — In some cases the loss of heat associated with and due to liquefaction is very great. Thus, when ice and salt are mixed, the ice melts and the salt dissolves in the water so formed. Thus both liquefy. The amount of heat absorbed from surrounding objects is very great, and the cold so produced is utilized in many operations in the arts.

The *mixture* remains liquid at temperatures much below that at which both constituents when separate would be solid. This clearly shows that the liquefaction is not due to heat alone, but involves also some specific influence of adhesion or chemical union or both together. It is of similar nature to the phenomena already mentioned under the title eutexia. (See p. 49.)

This topic has certain relations to the adhesion of solids to solids, but is more closely affiliated with the solution of solids in liquids.

CHAPTER XII.

THE ATTRACTIONS OF MOLECULES (*continued*).

II.—ADHESION (*continued*).

The study of the conditions under which solids dissolve in liquids naturally leads to a consideration of those under which solids may be separated again from liquids holding them in solution. But it is not intended here to extend the discussion to the formation of precipitates by sudden chemical reactions.

THE SEPARATION OF A SOLID FROM A LIQUID.

The comprehension of this subject may be facilitated by a few typical examples. These will develop the following simple but important principle: *As dissolving is favored by increase of quantity of solvent and by addition of heat, so separation of a solid from its solvent is favored by decrease of quantity of liquid and by decrease of heat.* The withdrawal of heat is almost always practicable; the decrease of quantity of solvent is practicable in cases of certain liquids, like water, carbon disulphide, alcohol, ether, and others *that easily evaporate.*

FIRST EXAMPLE.—*Cane Sugar.*

If a saturated aqueous solution of cane sugar *has some of its water removed by evaporation*, a portion of sugar corresponding to the amount of water so removed immediately settles out in the solid form. Incidentally this sugar forms crystals.

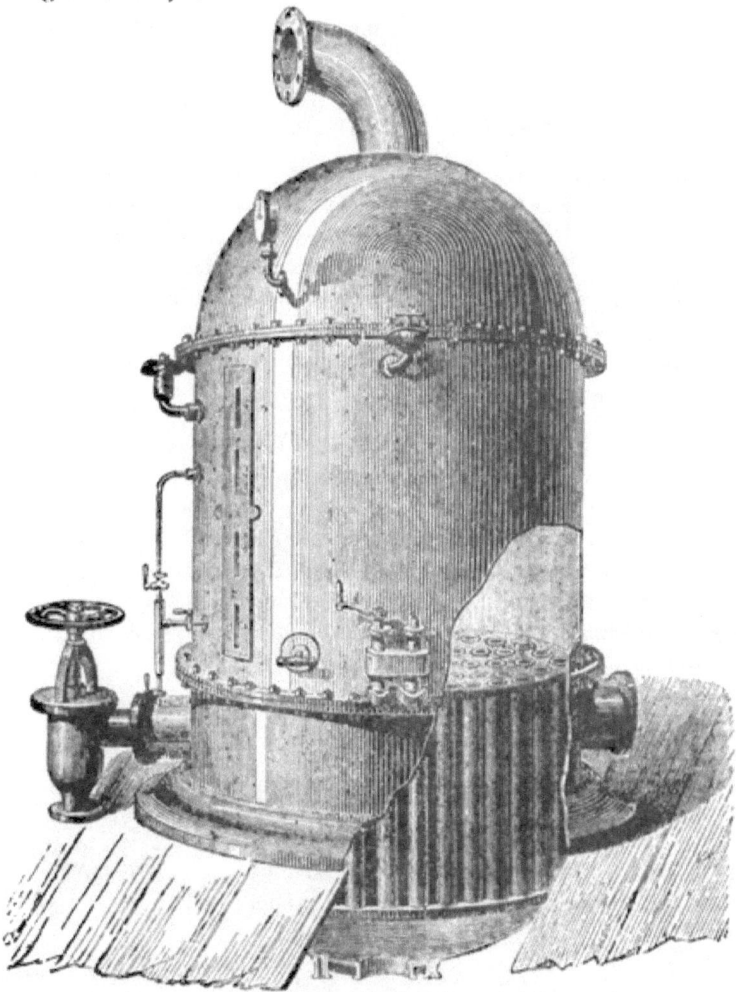

FIG. 81. — The vacuum-pan for producing rapid evaporation of water from sugar solutions. The vapor as fast as it is formed, and also the air in the apparatus, is rapidly withdrawn by a powerful pump. Whereupon further evaporation takes place, and the sugar syrup is brought to a condition such that it will readily crystallize.

Again, if a saturated aqueous solution of cane sugar *is reduced in temperature*, a portion of sugar corresponding to the amount of heat withdrawn immediately settles out in the solid form. In this case, also, the solidifying sugar incidentally crystallizes.

Both these means are, in fact, employed in the arts for the manufacture of sugar on the large scale.

Second Example. — *Alum.*

Potassic sulphate (K_2SO_4) and aluminic sulphate ($Al_2[SO_4]_3$) may be dissolved together in water. When the clear solution so formed is either evaporated or cooled, crystals of a new double salt separate. This salt is called alum. It is potassio-aluminic sulphate, and the crystals are found to have the composition represented by the formula

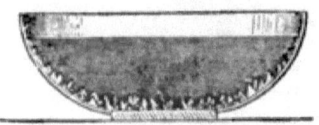

Fig. 82. — Crystals produced in a vessel by the slow evaporation of a liquid produced by solution.

$K_2SO_4, Al_2(SO_4)_3 \cdot 24 H_2O.$

Under these circumstances, then, the salts have the power of drawing to themselves, by reason of chemical affinity for it, a definite amount of water, called in this case, and similar ones, water of crystallization. Many other salts have the power of combining chemically with water in this way. In fact, it is believed to be probable that all substances that dissolve in water combine with it chemically, though the demonstration of the fact of combination sometimes involves difficulties.

Third Example. — *Sodic Sulphate.*

Sodic sulphate presents a peculiar and interesting form of the same tendency manifested by alum; *i.e.* to combine with water under proper conditions.

Thus there exist three salts — the same in composition except as respects water — differing merely according to the circumstances under which they solidify.

Fig. 83. — Burnt alum. When alum is heated in a crucible, it puffs up on account of the escape of water of crystallization in the form of steam.

These salts are: —

Anhydrous sodic sulphate	$Na_2SO_4.$
Heptahydrated sodic sulphate	$Na_2SO_4 \cdot 7 H_2O.$
Dekahydrated sodic sulphate	$Na_2SO_4 \cdot 10 H_2O.$

It is not necessary to undertake a detailed account of the conditions under which these several compounds are produced. It is sufficient to state, in general, that the formation of one rather than another is a matter of temperature mainly. The general rule is that in solutions of lower temperatures more water combines with the salt; in solutions of higher

FIG. 84. — Dr. Frederick Guthrie, lately Professor of Physics at the Royal School of Mines, London. Born in 1833; died in 1886.

temperatures a kind of dissociation takes place, and crystals are formed containing less water.

FOURTH EXAMPLE. — *Sodic Chloride.*

Common salt affords an illustration of the general principle just illustrated by sodic sulphate, only in the case of common salt the principle is extended to very low temperatures indeed. Crystals of common salt formed at ordinary temperatures are anhydrous; they have the formula NaCl. When a suitable solution is cooled considerably below the freezing-point of water, two kinds of crystals may be formed — one variety having the formula $NaCl \cdot 2H_2O$; the other variety, formed at still lower temperatures, having the formula $NaCl \cdot 10\tfrac{1}{2}H_2O$. Crystals formed in this way below the freezing-point of water are called, by Guthrie, *cryohydrates*.

The properties of the cryohydrates of common salt help to explain the well-known fact that salt water does not freeze except at temperatures much below $32°$ F. Salt water may be considered as a special chemical compound, the cryohydrate of common salt. This cryohydrate is characterized by a melting-point (and, what is the same thing, a solidifying-point) which happens to be below $0°$ C.

Efflorescence. — Most crystals containing water of crystallization may give it off by mere influence of heating. The amount of heat required varies with the substances: in some cases the ordinary heat of the atmosphere is sufficient. The result of such expulsion is a breaking of the crystals into a non-crystalline powder. The crystals are said to effloresce

Fifth Example. — *Metallic Lead.*

Melted lead is of course a liquid; when it is slowly cooled, it permits the formation of solid crystals. Many other melted metals and alloys do the same, but melted lead affords a good example, because in many large lead works this crystallization is continually carried on on an enormous scale. Lead, as produced from the ore, contains a minute amount of silver diffused through it. By Pattinson's process for extraction of this silver the melted mass is cooled and thus partly crystallized; solid crystals of nearly pure lead thus separate, and upon their removal they are found to have left most of their silver in the melted portion of lead remaining. From this the silver is finally extracted.

Alloys. — The term *alloy* was originally applied to a mixture of gold and silver melted together with or without other metals. The term is now applicable to all mixtures or compounds of metals with each other, except those containing mercury, which latter are called "amalgams."

On melting two metals together, complete assimila-

FIG. 85. — Pattinson furnace for separating crystals of pure lead (from its solution in melted argentiferous lead) by slow cooling.

tion takes place in some cases; in others it does not. Thus, silver does not readily alloy with iron.

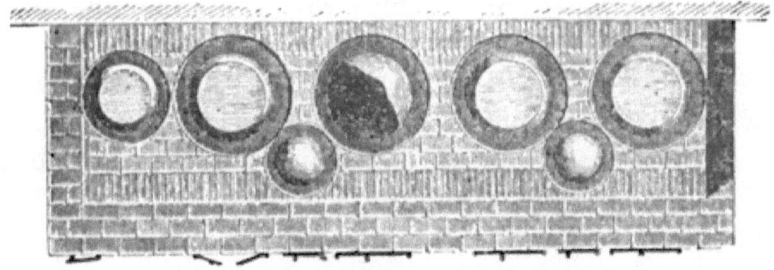

FIG. 86. — Top view of Pattinson furnace, showing the kettles in which argentiferous lead is melted.

The physical properties of an alloy are, in certain cases, the mean of the properties of the metals of which it is composed; in other cases they are widely different.

Matthiessen has divided the metals that form alloys into two classes:—

FIRST. Those which impart to their alloys their own properties: lead, tin, zinc, and cadmium.

SECOND. Those which do not: the other metals.

He regards the alloys of *class first* as solidified solutions of one metal in the other. The metals of *class second* he considers enter into alloys in allotropic form.

(*C*) ADHESION BETWEEN SOLIDS AND GASES.[1]

A solid, when immersed in a gas and then withdrawn, retains upon its surface a thin film of gas, somewhat as a solid is wetted by dipping in water.

Further, some solids absorb into their intermolecular spaces a great bulk of gas — so much indeed that in some cases the absorbed gas occupies a volume even smaller than it would if condensed to the liquid state by itself.

The metal palladium is remarkable for absorbing or *occluding*, at ordinary temperature, eight hundred times its bulk of hydrogen gas. The late Professor Graham, of London, who observed this property of palladium, considered the solid thus formed to be an alloy, and to contain hydrogen in the solid form. The quantities of the two elements are in this case approximately in the proportion of the weight of one atom of hydrogen to one atom of palladium, so that it has been suggested that the substances may be in chemical combination.

The metal platinum has the same power as palladium, though to a less degree.

If a warm piece of platinum foil is placed in a current of mixed illuminating gas and air, the foil absorbs portions of all the gases. In so doing it condenses them to such a degree as to bring the molecules very near to each other, even within that minute distance through which chemical

---

[1] Section (*B*) is at p. 114.

attraction can be exerted. Chemical union does in fact take place, as is evidenced by the production of light and heat and other phenomena of true combustion.

Hannay and Hogarth have shown that in some cases a gas, brought in contact with a solid, dissolves the latter quickly into itself.

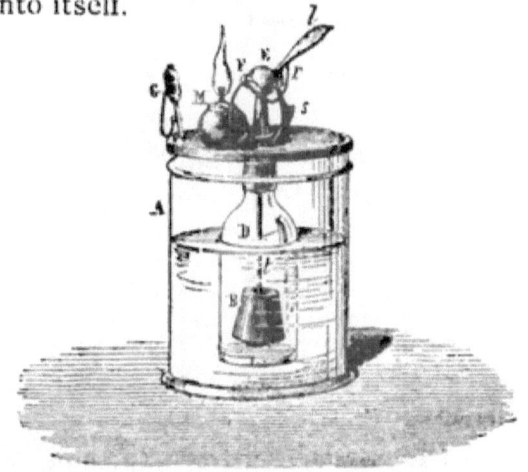

FIG. 87.—Döbereiner's lamp, showing the adhesion of hydrogen gas to platinum. The bottom of the lamp is a hydrogen generator. Dilute sulphuric acid acts upon the mass of zinc *B*. Hydrogen rises in the little bell-glass, and streaming from the tip at *F*, and falling upon the mass of spongy platinum at *G*, takes fire. The burning hydrogen lights the oil lamp *M*.

(*D*) ADHESION BETWEEN LIQUIDS AND LIQUIDS.

In general, the adhesion of liquids to liquids so far exceeds their respective cohesive forces that the liquids may be mixed in all proportions.

In general, heat favors this sort of diffusion.

Thus water and ordinary alcohol, when mixed in any proportion whatever, mingle throughout by virtue of their own attractive forces. On the other hand, when water and ordinary ether are mixed, only a certain small amount of the ether dissolves in the water; the ether in excess of this amount forms a separate layer upon the top of the water.

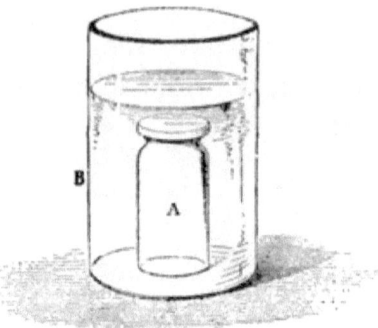

FIG. 88. — Apparatus for demonstrating the fact and the amount of liquid diffusion. A given liquid is placed in the vessel B. A solution to be tested is placed in the vessel A, provided with a glass cover. At a certain point of time the cover of A is removed. The material in A at once commences to diffuse into the liquid B. After a proper period of time has elapsed, the cover is replaced upon A. A portion of the liquid B is then tested, and the amount of material that has diffused from A into B in the given number of minutes or hours is determined.

There are many well-known examples of two liquids which scarcely mix at all; water and oil, water and mercury, are such.

FIG. 89. — Graham's apparatus for dialysis.

Osmose of Liquids. — In case of two liquids separated by a porous septum (it being granted that there exists adhesion between the liquids, and a difference in the amounts of adhesion of the two liquids for the septum), the liquid which wets the septum the better passes through the more rapidly.

Fig. 90. — Dialyzing apparatus separated. The upper vessel is called the dialyzer. It consists of a ring open at top and bottom, the bottom opening being covered with a membranous material, held in place by a stout rubber ring.

Dialysis. — The process of dialysis can be displayed by means of a suitable vessel divided by a kind of membranous partition into two com-

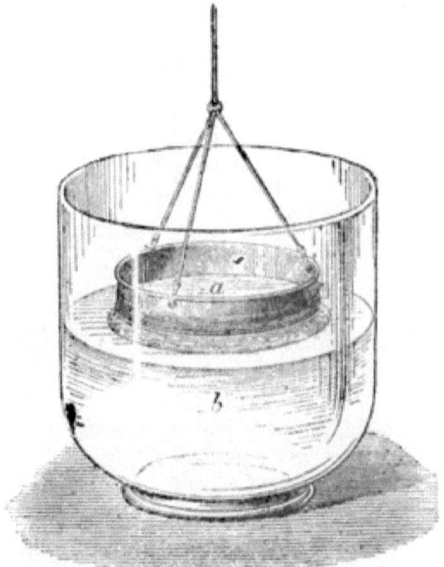

Fig. 91. — Graham's apparatus for illustrating dialysis. A crystallizable substance placed in the vessel a, called the dialyzer, passes by liquid diffusion into the liquid b.

partments. If pure water is placed in one compartment and the aqueous solution of some crystallizable substance in the other, dialysis takes place; that is, the crystallizable substance makes its way through the diaphragm into the other compartment. Non-crystallizable substances (for this purpose called *colloids*) are not capable of this kind of transfer. Evidently the crystallizable substances pass through the diaphragm by a kind of osmose.

(*E*) Adhesion between Liquids and Gases.

Water and many other liquids have the power of dissolving gases, though in very different proportions.

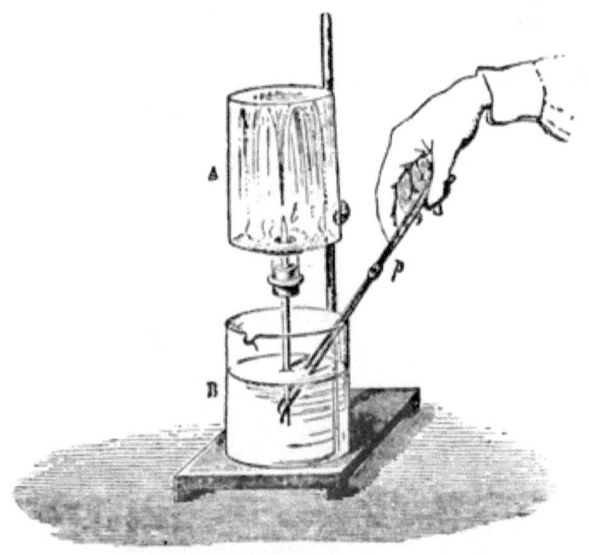

Fig. 92. — Ammonia fountain. The vessel *A* contains at first ammonia gas. As water from *B* passes up through the little tube, the ammonia gas dissolves so rapidly in the water as to produce diminished pressure. Whereupon the atmospheric pressure upon the surface of water in *B* forces the water into the vessel *A* as in a fountain.

The amount of gas absorbed by a liquid upon which it exerts no chemical action depends upon —

I. The nature of the gas and liquid;

II. The pressure to which they are exposed (the amount of gas absorbed varies directly as the pressure);

III. The temperature (with few exceptions, the solubility of a gas in a liquid is greater, the lower the temperature).

Evidences of the difference in the amounts of gas dissolved by a stated amount of water may be found in the following table: —

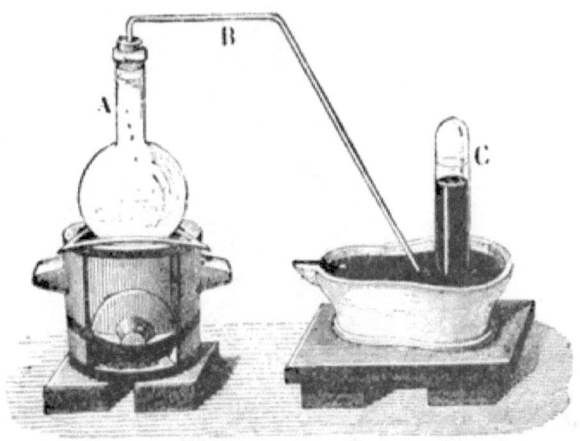

FIG. 93. — Disposition of apparatus for showing the adhesion of atmospheric air to water. Water containing air is placed in flask *A*. Upon boiling this water air is expelled and some steam is formed. The steam and air pass into the bell-glass *C*. The air collects at the top of the bell-glass. The water-vapor condenses on the surface of the mercury.

TABLE

SHOWING AMOUNTS, BY VOLUME, OF SEVERAL GASES SPECIFIED, DISSOLVED BY 1000 VOLUMES OF WATER AT 32° F.

Amount of water used,			1,000 volumes.	
"	" hydrogen gas	dissolved,	19	"
"	" nitrogen gas	"	20	"
"	" oxygen gas	"	41	"
"	" carbon dioxide gas (CO_2)	"	1,796	"
"	" hydrosulphuric acid gas (H_2S)	"	4,370	"
"	" sulphur dioxide gas (SO_2)	"	68,861	"
"	" ammonia gas (NH_3)	"	1,049,600	"

The very large amounts in several of these cases are believed to be due to the definite chemical union of the gases with the water to form new compounds.

It is a fact worthy of mention that molten silver has the power of drawing oxygen from the air and dissolving it in a quantity equal to twenty

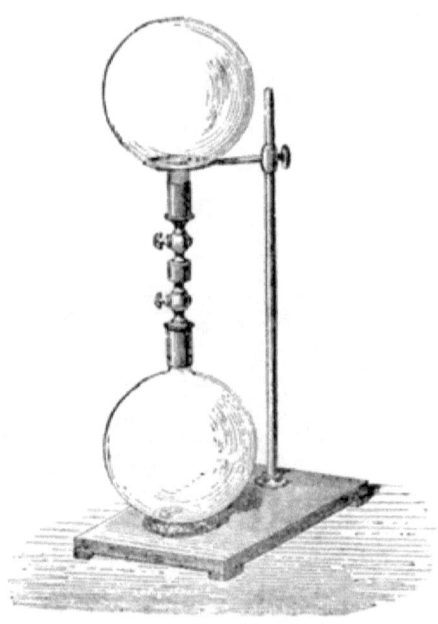

FIG. 94. — Apparatus for illustrating diffusion of gases. If a heavier gas is placed in the lower flask and a lighter gas is placed in the upper flask, and the stop-cocks are opened, it is found experimentally that the lighter gas diffuses rapidly downward into the other, and that the heavier gas diffuses upward (although more slowly) into the lighter.

times its own volume. When the silver solidifies, this gas is violently expelled. (The same principle is manifested by water; upon freezing, it expels the oxygen and nitrogen it previously dissolved from the air.)

(*F*) Adhesion between Gases and Gases.

The extraordinary tendency of gases to intermingle and interdiffuse has already been discussed under the title of diffusion of gases. This tendency is so strong that it overcomes the greatest differences of specific gravity.

These phenomena are not mainly due to adhesion, however, though there are grounds for believing that there is such a thing as gaseous adhesion. Thus Regnault has shown that when a liquid evaporates in the air, more vapor rises than when it evaporates into the same volume of vacuous space.

The tendency of gases to intermingle seems to be mainly a development of their tension or expansive power. This phenomenon is due to that motion within the mass which the molecules of all kinds of matter — even the most rigid — possess.

But the molecules of the gaseous form of matter are almost uninfluenced by *cohesion*. Hence they *manifest* this intermolecular motion to the most striking degree. And so when gases themselves are compared, it can be proved that the molecules of the lightest ones move with the greatest rapidity. Of course the ample spaces between the molecules of a gas offer great opportunities for the entrance of the molecules of another gas.

The Terrestrial Atmosphere. — The atmosphere of our globe affords a splendid example of gaseous diffusion constantly at work on a large scale.

1. The atmospheric air consists mainly of a mixture of oxygen gas and nitrogen gas, in the following proportions : —

COMPOSITION OF ATMOSPHERIC AIR.

	By volume.	*By weight.*
Oxygen	20.9 per cent.	23.1 per cent.
Nitrogen	79.1 "	76.9 "
	100	100

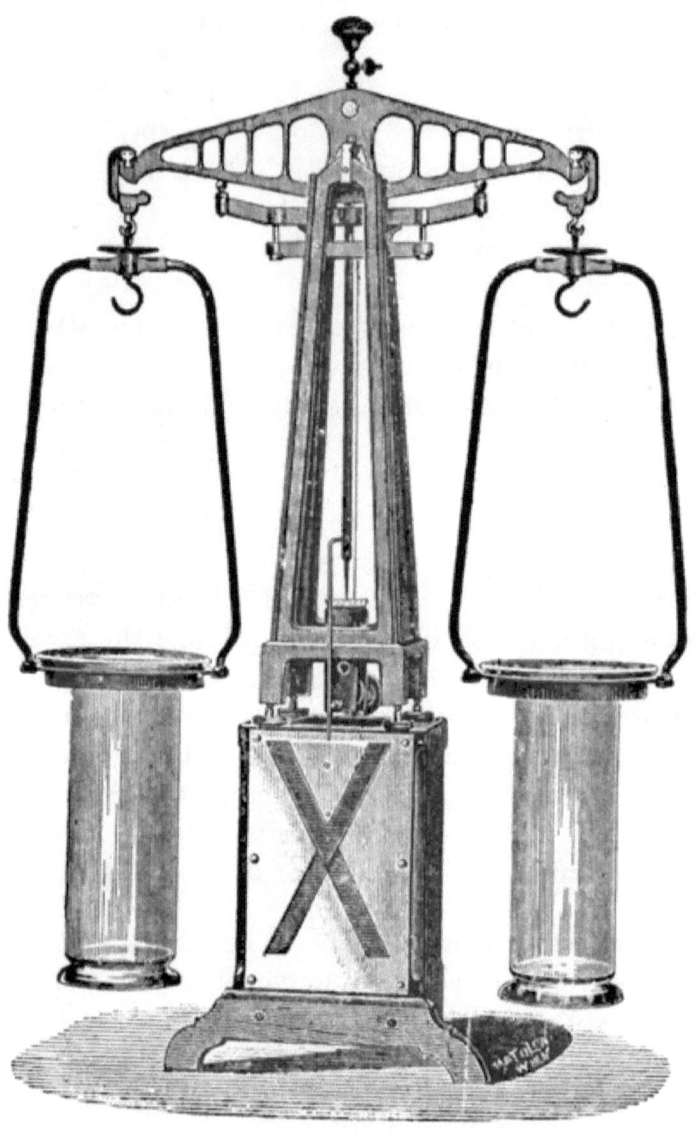

FIG. 95.—Balance for showing that certain gases are heavier than the atmosphere. The one jar contains atmospheric air. When a heavier gas is poured into the other jar, the needle of the balance is boldly deflected.

THE ATTRACTIONS OF MOLECULES.

Now any given measure of oxygen gas is sixteen times as heavy as the same measure of the standard gas, hydrogen; but nitrogen gas is only fourteen times as heavy as hydrogen. Yet in our atmosphere the heavier oxygen does not settle out, but remains thoroughly intermingled with the nitrogen.

2. The respiration of living animals and the burning of all our chief fuels are constantly casting into the atmosphere immense quantities of a heavy gas, carbon dioxide (CO_2). This gas is twenty-two times as heavy as the standard gas, hydrogen. Of course, therefore, it is much heavier than the oxygen or the nitrogen of the atmospheric air; it does not settle out from the air, however, but promptly intermingles with it and remains intermingled.

FIG. 96. — Regnault's method of suspending, from the balance-pan, a globe containing a gas to be weighed. A globe of similar volume is also suspended from the other pan.

NOTE I. On the density of atmospheric air.

The air contains minute amounts of a multitude of gases, but oxygen and nitrogen so largely predominate that only these need be taken into the account here.

The density of the air is somewhere between the densities, 16 and 14, of its two chief constituents: it is about 14.4.

1 volume of oxygen gas, weighing 16 units	16. units.
4 volumes of nitrogen gas, each weighing 14 units . .	56. "
5 volumes of mixture (air) will weigh	72. "
1 volume of air will weigh	14.4 "

NOTE II. On the density of carbon dioxide gas (CO_2).

By actual weighing, in comparison with an equal volume of the standard gas, hydrogen, this gas has been found to have the density 22; *i.e.* to weigh, bulk for bulk, 22 times as much as hydrogen.

The density may be computed from the molecular weight as follows: —

Formula of a Molecule of Carbon Dioxide Gas (CO_2).

Weight of one atom of carbon . . . . 12 microcriths.
"   " two atoms of oxygen (16×2) . . . 32   "
"   " one molecule of carbon dioxide . . 44   "
"   " one molecule of hydrogen, H_2 (1×2)   2   "

Hence a molecule of carbon dioxide weighs twenty-two times as much as a molecule of hydrogen.

But all gaseous molecules have the same size; hence, any volume of carbon dioxide weighs twenty-two times as much as the same volume of hydrogen.

NOTE III. Of course, in weighing atmospheric air and other gases, pressure and temperature must be considered. The pressure must be measured by some form of barometer. The temperature must be measured by some form of thermometer.

CHAPTER XIII.

THE ATTRACTION OF ATOMS.

CHEMICAL AFFINITY.

CHEMICAL affinity is an agency which acts at insensibly small distances, and tends to produce combinations of certain atoms and molecules of matter into groups of a precisely determinate kind.

The characteristics of this agency cannot be described in a few words. To it are referred a multitude of phenomena, displaying under different circumstances the greatest variety of action. Such differences are for example: —

As to the original *quantity* and *intensity* of the activity itself.

As to the *conditions* under which its active powers are displayed.

As to the *methods* by which it works.

As to the *sphere of activity* — extremely narrow in a certain sense and extremely wide in another.

As to the *results* accomplished by it.

The Conditions favoring Chemical Change. — 1. This force manifests its chief activity between atoms or molecules *of different kinds.*

Thus, an atom of hydrogen has affinity for another atom of hydrogen, and the two may unite to form a molecule of hydrogen, expressible by H—H, also written H_2. Again, an atom of chlorine has affinity for another atom of chlorine, and these two may unite to form a molecule of chlorine expressible by the formula Cl—Cl, or Cl_2. But when a molecule of hydrogen is brought in contact with a molecule of chlorine, the two generally suffer decomposition, so that a rearrangement may take place and two new molecules of hydrochloric acid (HCl) may be produced. This chemical change may be expressed by the following equation:—

$$H_2 + Cl_2 = 2\,HCl$$

H_2	+	Cl_2	=	$2\,HCl$
One molecule of Hydrogen,		One molecule of Chlorine,		Two molecules of Hydrochloric acid,
2 parts by weight.		71 parts by weight.		73 parts by weight.
	73			73

Evidently the atom of chlorine has more affinity for an atom of hydrogen than for another atom of chlorine. And an atom of hydrogen has more affinity for an atom of chlorine than for another atom of hydrogen.

2. It is manifested between different substances with very *different*, though *definite*, degrees of force. Thus the metal gold and the metal iron oxidize (that is, combine with oxygen) with different degrees of ease; but it is always the iron that oxidizes the easier.

3. Certain *physical conditions* are of great importance in connection with chemical action. When physical conditions are favorable, chemical action proceeds with great vigor; when they are unfavorable, the same processes sometimes fail to advance at all, or they may be even reversed. Under unfavorable conditions chemical affinity appears either not to exist or to be dormant.

The following are some of the physical conditions which determine chemical changes — changes that may have as their prominent features either the building up or the breaking down of molecules: —

(*a*) **The Liquid Condition.** — Some substances that chemically unite when mixed as solutions, manifest no affinity when they are mingled in the solid form. Thus, solid tartaric acid and solid hydro-sodic carbonate when mingled manifest no change. When water is added, however, each solid dissolves, and a chemical change at once ensues, hydro-sodic tartrate, carbon dioxide, and water being formed.

The chemical change may be expressed as follows: —

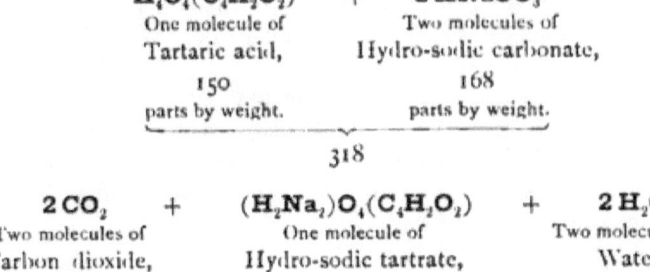

The equation indicates that water is actually formed by the operation; it appears evident, therefore, that the water which acted as the solvent was not demanded in the building up of the molecules produced, but did, in fact, act as a favoring physical agent.

It appears to be proved, however, that certain solid bodies, finely pulverized, thoroughly mixed, and then subjected to great pressure, produce new compounds as the result of the pressure (and not of the heat attendant). The quantities of the compounds changed appear to increase with the *duration* of the pressure and its *amount*, as well as the fineness and thoroughness of intermingling of the powders.

Thus, in a certain experiment, mixtures of dry, pure precipitated baric sulphate and sodic carbonate were subjected to a pressure of six thousand atmospheres under varying conditions of temperature and duration of the pressure. Afterward the product was tested. After a single compression the amount of baric carbonate produced was about one per cent; the solid block produced was pulverized and compressed again, when five per cent of barium carbonate was produced; further treatment brought it up to eleven per cent.

It has been concluded that —

1. A sort of diffusion takes place in solid bodies.
2. Matter assumes under pressure a condition relative to the volume it is obliged to occupy.
3. For the solid state, as for the gaseous, there is a critical temperature above or below which changes by simple pressure are no longer possible.

(*b*) **Heat**. — Many substances, when practically in contact with each other, do not combine chemically unless the whole or a portion of the mass is raised to some definite point of temperature. When this point is reached, union at once commences.

The process of combustion of ordinary fuels affords an appropriate illustration. If a portion of a mass of coal is heated in the air to the point at which union with oxygen takes place, the phenomena of combustion (a form of chemical union) are witnessed. The chemical change initiated may be expressed in part as follows: —

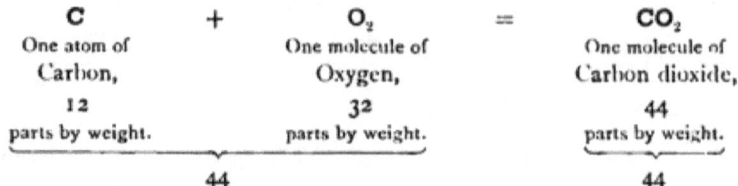

It is an interesting fact that generally the combustion of the first portions of the coal evolve, by the act of chemical union, sufficient heat to raise yet other portions to the igniting point. This process, repeated, enables the operation to proceed from portion to portion so long as the supply of carbon and oxygen are kept up — unless, indeed, some unfavorable physical condition is allowed to supervene.

There are numerous other examples known, in which chemical action is stimulated by an amount of heat insufficient to produce light.

In fact, addition of heat is the method oftenest used for developing or arousing chemical affinity.

Thermolysis and Dissociation. — Another, and at first seemingly inconsistent chemical effect of heat, ought to be mentioned here. It has already been pointed out that addition of heat expands material bodies, and even changes solids and liquids to the gaseous form. (See p. 43.) These effects are believed to be essentially associated with a motion of the particles of the body, such that the molecules are moved farther and farther apart, and even beyond the range of influence of those cohesive forces

FIG. 97. — Henri St. Clair Deville, distinguished French chemist, noted for his discoveries in the chemistry of high temperatures, dissociation, for example.

that bind them into solid and liquid masses. It would be quite consistent with this view if still greater addition of heat were found to be sufficient to drive even atoms apart from each other, and so to place them beyond the minute distances within which the force of chemical affinity is exerted. This would result in a decomposition of *compound* molecules and a lessening of the number of atoms capable of existing together in elementary molecules.

Now the experiments of Deville and others fully confirm these suggestions. It is, in fact, proved that certain substances, as water, for example, may be decomposed into their elements by influence of high temperature alone. In this, and some similar cases, the elements may reunite when the temperature of the mixture falls slightly. This kind of temporary decomposition is called *dissociation*. It may be added that light and electricity, as well as heat, are in some cases capable of accomplishing it.

In another class of cases, of which ammonia gas (NH_3) may serve as an example, the molecule is *permanently* broken up; that is, its elementary substances do not, by fall of temperature, rejoin to produce the original compound. In such cases the operation is called *thermolysis*.

It is also observed that certain elementary substances, as sulphur, for example, manifest a gradual lessening of their relative vapor densities as they are raised to higher and higher temperatures. This lessening of vapor density is accepted as an indication that the molecules contain fewer and fewer atoms; that is, undergo dissociation. The methods of Victor Meyer and others have directed attention to this subject.

Heat often produces a modification of the *relative* chemical attractions of bodies. Thus, at ordinary temperatures, sulphuric acid is capable of displacing boric acid from its salts in solutions. At high temperatures — the red heat, for example — the chemical affinities are reversed: boric acid displaces sulphuric acid.

In a few cases chemical decomposition is producible by mechanical means, as, for example, in certain explosive compounds; but it is probable that the mechanical is not always the immediate cause. In the familiar cases where mechanical percussion produces decomposition of certain explosives, evidently *the heat* generated by the percussion is the true cause.

(*c*) **Light**. — This agent, as usually produced by luminous bodies, is by no means a homogeneous one; the prism shows it to be divisible into thousands of kinds of energy, characterized by greater or less differences. The white light, as emitted by most of its sources, has at least three classes of rays, — luminous rays of various colors, non-luminous chemical rays, non-luminous heat rays. The non-luminous chemical rays, called also actinic rays, have a specific power of determining the chemical union of certain elements and the chemical decomposition of certain compounds.

Thus chlorine gas and hydrogen gas, when mixed in a dark room, do not readily unite; when such a mixture is exposed to sunlight, almost instantaneous combination ensues.

The decomposing influence of certain rays of light is displayed in the photographic print, the substance decomposed being argentic chloride.

(*d*) **Electricity.** — The influence of electricity in connection with chemical action is manifested in at least four different forms.

FIRST. Operations coming under the head of *electrolysis*.

Electricity of low tension, such as that produced by the galvanic battery, is capable of most important influences on chemical compounds. In

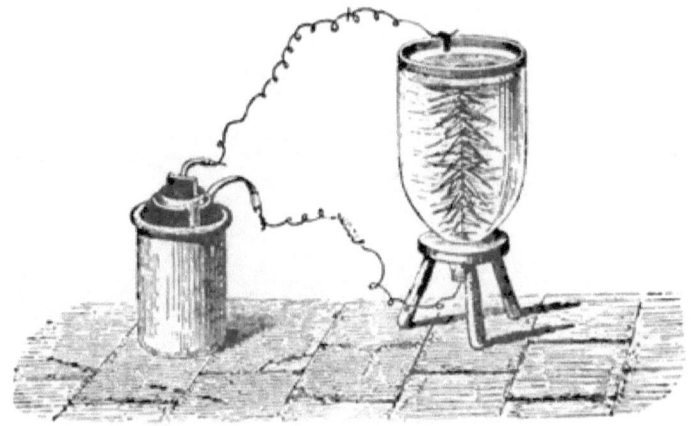

FIG. 98. — Apparatus showing how an electric current may precipitate a metal from its solution.

processes of electro-plating with copper, nickel, silver, gold, and other metals, it sets in motion an invisible current by which atoms of metal are driven away from the metallic plate called the *anode*, then into the molecules of acid or metallic salt dissolved in the plating-bath, and thence upon the surface of the object to be plated, called the *cathode*. The *deposition of the metal* is merely incidental; the *molecular transfer* is the important feature. Metals may be dissolved by this method if desired. Again, non-metals may be made to combine, or, if combined, may be separated in like fashion by the current.

SECOND. When a current, in the form of an *electric arc*, passes through compounds or through mixtures of elementary substances, chemical changes often occur. One of the most marked illustrations of this kind of influence is in the direct union of carbon and hydrogen whereby the substance known as acetylene (C_2H_2) is formed. One of the most important features of interest in connection with this operation is the fact that acetylene represents a starting-point for the synthesis, or building up, of organic compounds directly from their elements.

THIRD. Another form of action is by the influence of the *electric spark* from a Ruhmkorff coil, a Holtz machine, or similar appliance. In this way distinct chemical action is stimulated over a limited field. The field, however, may be widened by continuance of the electric discharge.

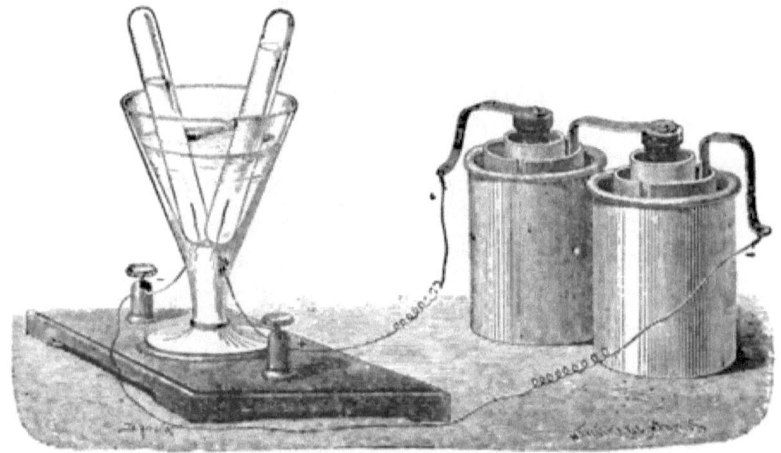

FIG. 99. — Apparatus for decomposing water into its components, hydrogen and oxygen, by means of a galvanic current generated by a Bunsen battery.

Such an electric discharge, in the form of sparks flowing from platinum terminals through dry atmospheric air, gives rise to a direct union of the oxygen and nitrogen. Brown fumes of N_2O_4 or NO_2 are thus formed. These with water may form nitric acid (HNO_3).

Probably lightning discharges form nitric acid, in this way, and thus contribute to the available nitrogen of the soil.

FOURTH. The *silent electric discharge* produces certain marked effects, of which the most noteworthy is the change of ordinary oxygen into ozone.

In all these cases the action may be twofold. There is the true electric influence, and, especially in cases of the second and third methods already referred to, there is the additional influence of the heat connected with the arc or with the luminous discharge.

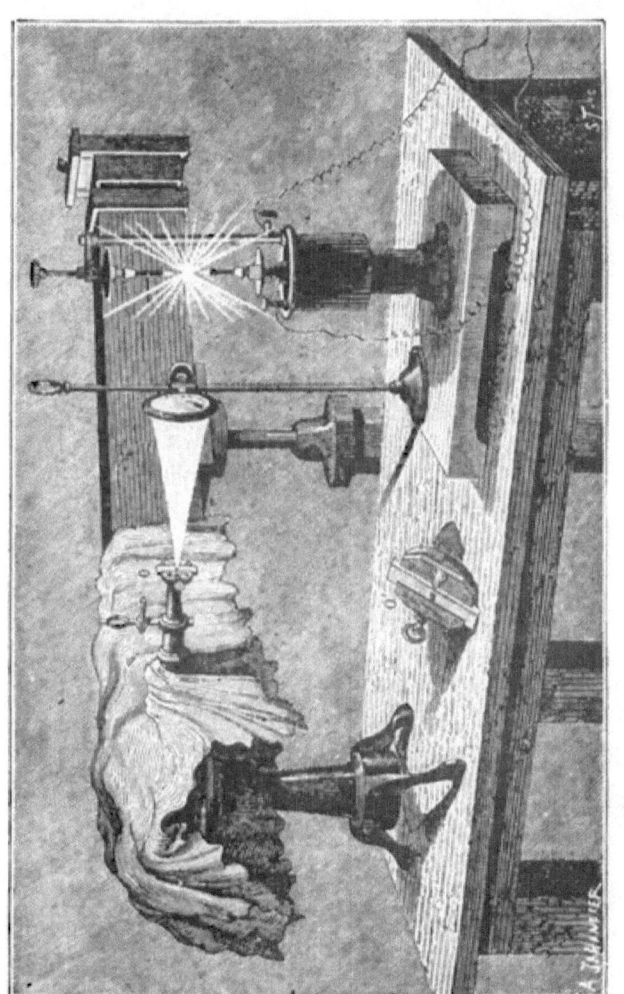

FIG. 100. — Lockyer's apparatus for analysis of alloys of gold and silver by use of an electric current, a spectroscope, and a photographic camera.

150   THE ATTRACTION OF ATOMS.

(*e*) **Vital Processes of the Higher and Lower Living Beings.** — The vital powers of the *higher orders of animal and vegetable beings* have most marked influence upon chemical action. Thus vast numbers of compounds are recognized as existing in living animals and plants that have not yet been produced without the intervention of vital force. A few crys-

FIG. 101. — Ruhmkorff, the celebrated manufacturer of electric instruments, and inventor of the Ruhmkorff coil.

tallizable substances, ordinarily the products of living organisms, have lately been produced by circuitous chemical operations without intervention of life. Doubtless others will be in future.

Certain processes of acetic, butyric, and other fermentations, purely chemical in their nature, have been shown to be due to the presence and action of *microbes*, and not to go on in their absence.

THE ATTRACTION OF ATOMS. 151

Organic and Inorganic Compounds. — Chemists long ago recognized certain differences between the substances found in distinctly animal and vegetable matters, on the one hand, and the substances found in mineral matters, on the other — between those things which constitute organisms like animals and plants, as opposed to non-living substances like clay, iron-rust, alum, saltpetre, etc.

Animal matters and vegetable matters are the products of bodies possessing organs. Organs are parts having specific functions. Thus the stomach is an organ possessing the function of digestion, and the lungs are

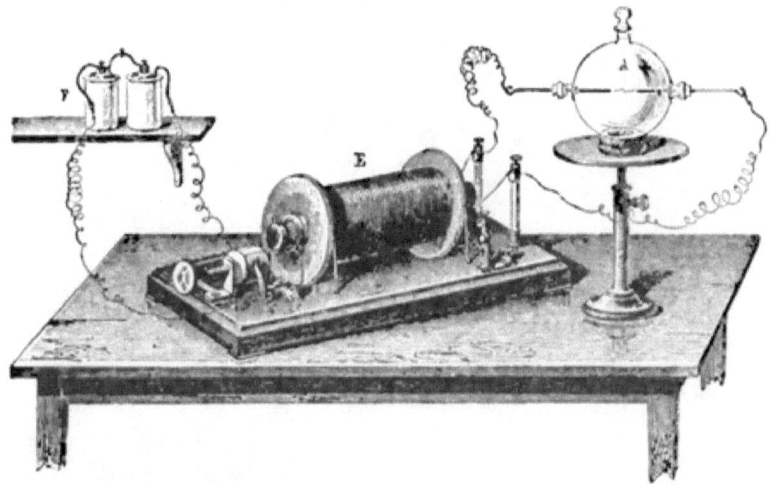

FIG. 102. — The Bunsen battery, F, energizes the Ruhmkorff coil, E, affording a series of electric sparks in the flask A. The flask contains a mixture of nitrogen and oxygen in a very dry condition. The sparks lead the two gases to combine.

organs possessing the function of respiration. Again, the leaves, the flowers, the seeds, the roots, of plants, are separate organs, and they possess special and very different functions of the living vegetable to which they belong. *Accordingly, substances derived from vegetables and animals are called organic.* Non-living objects, as rocks and other mineral and earthy substances, do not possess organs, and they have long been called inorganic.

This division of matters into organic and inorganic was formerly thought an essential one; it is not now considered so. It is now known that the chemical changes of living animals and plants are governed by the same laws as those prevailing in the changes of rocks and other lifeless forms of matter.

Grounds for this Division.— Chemistry is still, however, commonly divided into the two great departments, — inorganic chemistry and organic chemistry; but this division is recognized as a matter of convenience mainly.

FIG. 103. — Yeast plant, illustrating the formation of additional cells by fission.

Three reasons, which may be mentioned, why the distinction is still maintained, are : —

FIRST. The number of organic compounds is very great.

 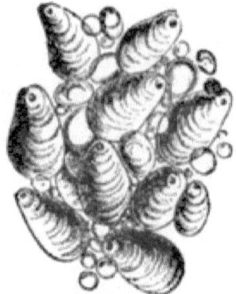

FIG. 104. — Globules of wheat starch, as seen under the microscope, showing cellular structure.

FIG. 105. — Globules of potato starch, as seen under the microscope, showing cellular structure.

SECOND. These compounds perform varied and important offices in connection with human beings in their growth and nourishment in health, and in their treatment in illness.

THIRD. The processes of analysis and the methods of investigation in organic compounds are slightly different, as a whole, from those that serve for the study of inorganic.

Definition of Organic Chemistry. — The inorganic and the organic worlds are, however, so closely allied in some respects, and certain of the substances of the one have such close and natural affiliations with those of the other, that it is often found difficult to determine where shall be placed the line of demarcation between these two great natural groups. In fact, chemists have not found the definition incidentally introduced in a preceding paragraph sufficiently distinct. To make it more so, organic chemistry has been sometimes called *the chemistry of the carbon compounds*. It has sometimes been called *the chemistry of the*

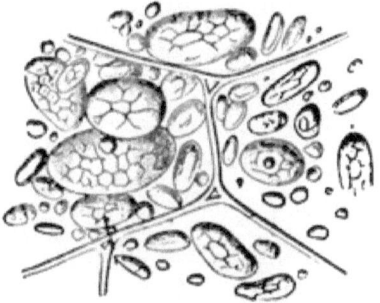

FIG. 106. — Globules of corn starch as seen under the microscope, showing cellular structure.

hydrocarbons. Again, the following still more rigid and scientific statement is often employed: organic chemistry includes those *compounds in which the atoms of carbon are directly united either with other atoms of carbon, or with atoms of hydrogen, or with atoms of nitrogen*.[1]

Two Classes of Organic Compounds. — There is one distinction between the classes of organic compounds themselves that ought not to be omitted here. The members of the organic family differ very much in their properties, according as they are *crystalline* or *cellular*. Crystalline organic compounds, of which cane sugar may be taken as a familiar and suitable example, are numerous. These compounds are closely allied in some respects to inorganic compounds. They do not seem to have so

---

[1] From the Author's work, "Beginner's Handbook of Chemistry."

close a relation to the vital processes as might at first be supposed. But those organic compounds that are cellular, such, for example, as the different varieties of starch, the fibre of wood, and the fibre of lean meat, are much removed from inorganic bodies, and seem to bear a peculiar and close relation to the vital forces. In general, cellular organic compounds are called *organized;* while the non-cellular organic compounds are called *non-organized.*

Compound bodies then are divided from a certain point of view into two great classes, — inorganic and organic. The organic are again divided into two classes, — organized and non-organized.

CHAPTER XIV.

THE ATTRACTION OF ATOMS (*continued*).

THE CHEMICAL WORK OF MICRO-ORGANISMS.

It is difficult to form a proper conception of the vast amount of chemical work accomplished by those ex-

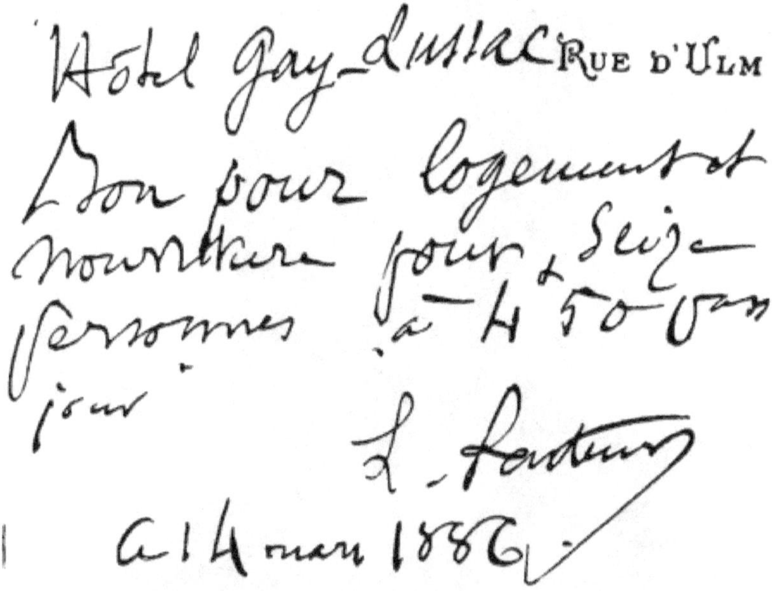
Hôtel Gay-Lussac Rue d'Ulm
Bon pour logement et nourriture pour seize personnes à 4 50 par jour
L. Pasteur
le 14 mars 1886

FIG. 107. — Autograph letter of Louis Pasteur, being an order for board for hydrophobia patients undergoing treatment at the Pasteur Institute.

ceedingly minute parasitic plants called micro-organisms, microbes, bacteria, etc. Of late years, following the work of Pasteur and Koch, many observers have industriously studied this subject. As a result, microbes or bacteria (the word *bacterium* is used in a general as well as in a special sense) are now recognized as of high importance in chemistry.

FIG. 108.—Dr. Robert Koch, celebrated investigator in the field of bacteriology.

These organisms are exceedingly numerous as varieties and yet more as individuals. They are most widely diffused. They exist in air (though not largely in sea air nor even in the air of large sewers), in water (though in varying quantities and kinds), in the soil (though in most cases, not at great depths).

They effect many kinds of decomposition of molecules — a work essentially chemical. Some of it is of industrial interest in manufactures and in agriculture; some of it leads to the wonderful fermentative and putrefactive processes of the world; some of it goes on as the chief factor in diseases of the higher animals, and indeed in the normal digestive operations of them.

FIG. 109. — Microscopic infusoria such as are found in stagnant natural water.

(Thus it is observed that not all microbes are pathogenic: some are distinctly beneficial to living animals.)

General Description of Microbes. — The organisms in question consist of very minute cells — often not greater in length than one ten-thousandth of an inch.

Yet, when floating in the air, they cannot pass through a small plug of loose cotton fibres.

In form they vary very much. The common forms are *the globular* (micrococcus), *the keyhole-shaped* (like two spheres in contact, or partly run together), *the rod-shaped* (bacillus form), *the comma-shaped, the spiral shaped*.

158   THE ATTRACTION OF ATOMS.

The cells are often grouped in a tolerably definite way in filaments or chains; sometimes they are gathered in great irregular masses.

A given cell usually consists of a sac of *mycoprotein* enclosing homogeneous protoplasm.

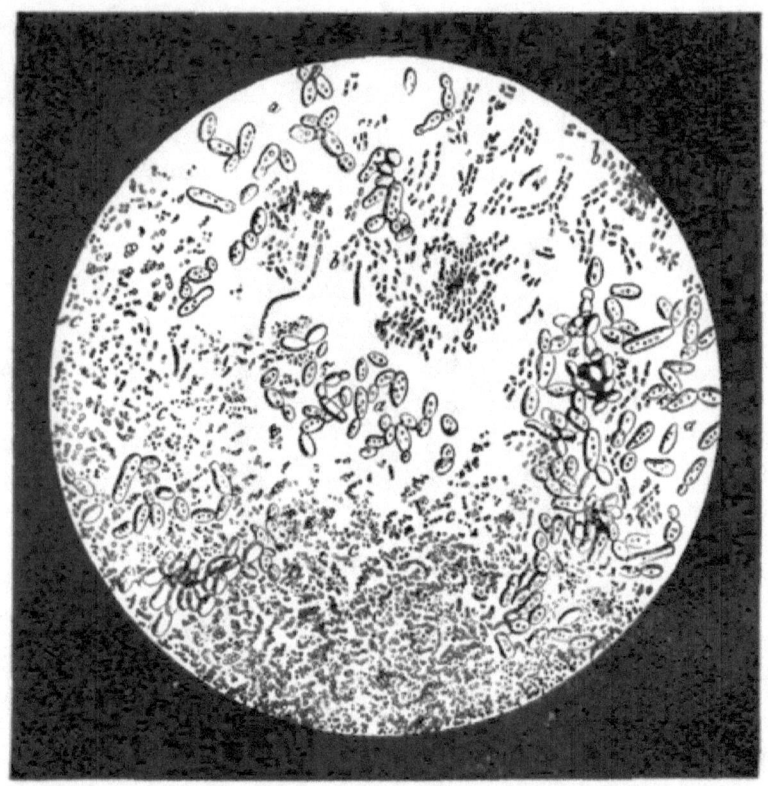

FIG. 110. — *a a*, mycoderma vini; *b b*, mycoderma aceti (earlier stage of development); *c c*, mycoderma aceti (advanced stage of development).

Classification and Nomenclature of Bacteria. — Micro-organisms have been divided into two sections, — (1) the *Endosporea* and (2) the *Arthrosporea*.

The former consists of but one genus, sporobacterium, which has four recognized species.

The latter (Arthrosporea) has two genera, bacterium and micrococcus.

The genus bacterium is the more numerous, having at least twenty-five distinct species, among which are the bacteria found in the human body in the diseases of consumption, pneumonia, cholera, diarrhœa, typhoid fever, and glanders; also the forms which are found in foul ponds and sewage: it also includes the vinegar ferment, which converts ethylic alcohol into acetic acid.

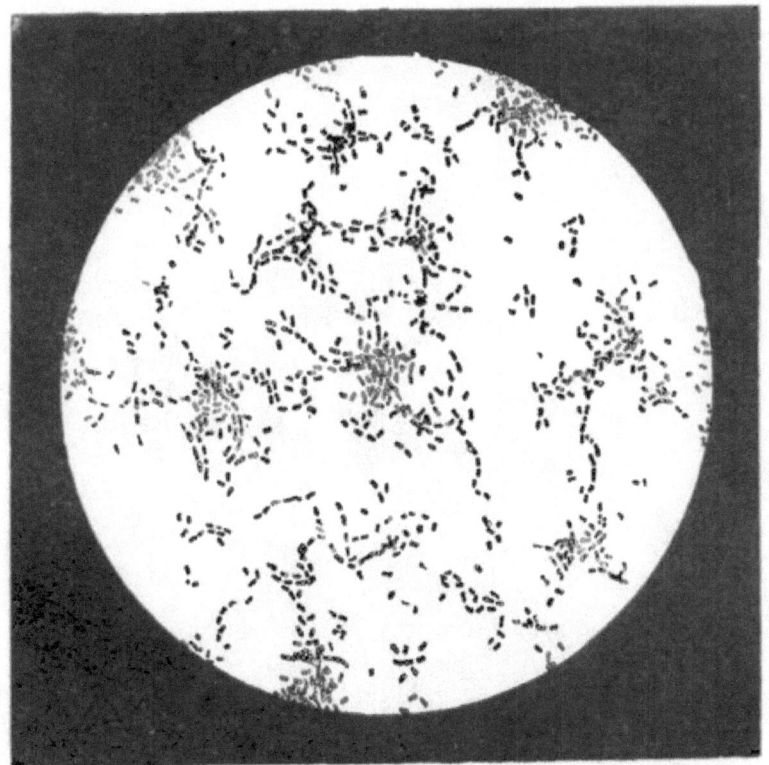

FIG. 111. — Mycoderma aceti, or mother of vinegar, as seen (enlarged 500 diameters) under the microscope.

The genus micrococcus has eight species now enumerated, among which are the bacteria of small-pox, erysipelas, scarlet fever, and others.

Growth of Microbes. — The cells of micro-organisms are capable of extremely rapid multiplication — generally by *fission* or some modification of it. Sometimes fission takes a course whereby forms like spores are produced. By such processes a single bacterium cell may multiply in twenty-four hours to more than a billion individuals like itself.

FIG. 112. — Apparatus used in Pasteur's laboratory in Paris; boiler (sterilizer) heated by gas, for destroying microbes by steam under high pressure; oven for culture of certain microbes at a definite temperature; oven for sterilizing tubes and flasks by hot air.

The number of microbes in some kinds of food is very great. Thus, it has been computed that in the case of certain kinds of Swiss cheese one pound of the article possesses a microbian population greater than the human population of the terrestrial globe.

FIG. 113.—Large oven used in Pasteur's laboratory for the culture of certain selected microbes. (It is provided with an automatic gas regulator.)

Micro-organisms are influenced in their growth by prevailing conditions; some conditions are highly favorable, some unfavorable.

1. *They need a certain temperature,* varying in particular cases (100° F. is generally favorable). They are best killed by very high temperatures

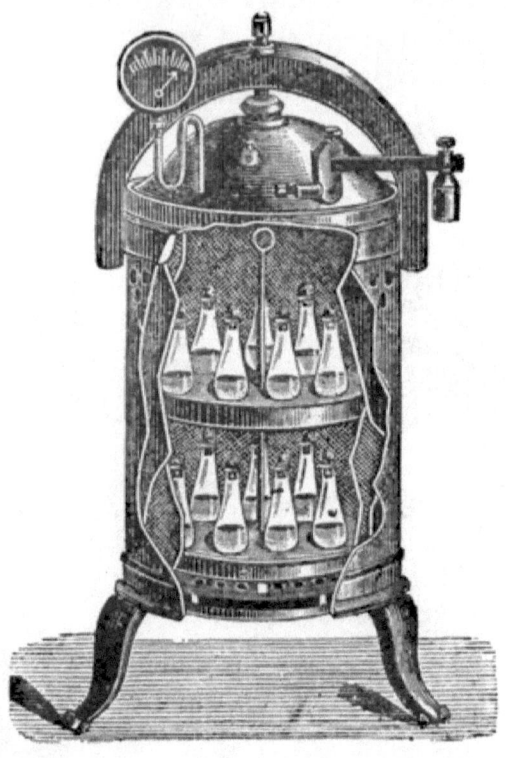

FIG. 114. — Sterilizing apparatus. It is for the purpose of destroying bacteria in such substances as may be placed in the flasks. The apparatus consists essentially of a strong vessel, made tight, and provided with a safety valve and steam gauge. It has jackets to prevent loss of heat and condensation of steam. Upon applying a strong gas flame underneath the vessel, the water in the vessel is raised to the boiling-point. High pressure steam is produced. The flasks are therefore subjected to a temperature sufficient to destroy any microbes in them. The flasks are then withdrawn all at once by means of the caster. The plugs of sterilized cotton in the necks of the flasks prevent subsequent access of microbes from the air.

(212° F. and upward). Some, however, succumb at even moderately low temperatures (50° F. and downward).

2. *They flourish best in presence of moisture.* A small amount will serve. Moderately dry dust containing them, when put under proper conditions with moisture, shows life by the multiplication of the varieties present. But *thorough* desiccation is fatal.

Fig. 115.—Louis Pasteur, celebrated for his studies in the diseases produced by micro-organisms.

3. *Some of them need atmospheric air;* some flourish best in absence of it.

As bacteria are destitute of chlorophyll, they do not obtain nutrition by decomposing carbon dioxide of the air under influence of sunlight.

4. *They must have suitable pabulum.* Farinaceous matters are good; albuminoid or other nitrogenous substances are very favorable; meat extract is excellent.

FIG. 116. — Burning sulphur and tar in the streets of Cairo, Egypt, for disinfecting purposes, during the prevalence of cholera.

THE ATTRACTION OF ATOMS. 165

Some bacteria attack only dead organized tissues; others attack and disorganize tissues of living beings.

Some obtain nitrogen from as simple compounds as ammonia gas (NH_3), others require albuminoid compounds. Some obtain carbon from substances such as ... others from tartaric acid; others require a more ... glycerin; yet others can take carbon as ... from proteids.

Thus it appears that, like animals, ... complex molecules formed by highly organized ...

5. *They are not materially interfered with by* ... *stances, even by some that are poisonous to* ...

But some chemical substances they do ... are the so-called germicides or antiseptics ... chloride, $HgCl_2$, is one of the most fatal ... used. The following are also more or less ... chlorine, zinc chloride, zinc sulphate, lead ... vitriol or copperas (ferrous sulphate), ... alcohol, carbolic acid (phenyl alcohol), and ...

In some cases chloroform suspends their ...

In many cases the growth of bacteria is ... own formation; these, when they are sufficiently ... poison the bacteria even though their habit ... ditions otherwise favorable to them are ...

6. *They must be free from interference* ... a large number of one kind of micro-organism ... smaller number of another kind.

But in some cases one species of bacteria ...

enable them to produce, in the aggregate, great quantities of such compounds as they are able to form.

The general tendencies of bacterial growth involve a *breaking-down of complex molecules* into somewhat simpler ones, although this is not the invariable result.

Microbes produce certain special compounds as has already been suggested.

1. They accomplish the following important transformations: —
Cane sugar to *ethyl alcohol;*
Glycerin to *ethyl alcohol*, and thence to *butyl alcohol;*
Cane sugar to *gum* or *mannite;*
Grape sugar or milk sugar or glycerin to *lactic acid* and *butyric acid;*
Urea to *ammonic carbonate;*
Hippuric acid to *benzoic acid;*
Albumens to *ptomaines;*
Nitrogenous matters to *nitrates.*

2. They produce certain groups of substances possessed of general properties, of which the following may be noted: —
Substances having marked agreeable or disagreeable odors;
Substances having brilliant colors;
Substances — called, in general, ptomaines — having eminently poisonous properties, as tyrotoxicon in milk and cheese.

The ptomaines just referred to are alkaloids of a highly poisonous character, generally resulting from a morbid decomposition of albuminoids under the influence of microbes.

The leucomaines are analogous poisonous alkaloids, but they are produced by the ordinary physiological processes of the higher animals, and thus are capable of being decomposed and excreted under the normal action of the appropriate organs, of which, apparently, the liver is the most effective.

It was formerly held that the morbid conditions recognized in animals affected by certain contagious and infectious diseases were due directly to the specific microbes present. At present the abnormal action of the organism is referred rather to the poisonous ptomaines produced by the microbes.

Usefulness of Bacteria in the Organic World. — One of the most marked features in the life-processes of the higher animals and plants is the circulation of certain atoms. That is, there seems to be a definite and rather small stock of certain useful elements, like nitrogen, phosphorus (and to these may be added, with less force, potassium and even carbon), which are in a continual state of transfer. This "stock" is absorbed from the soil by living plants; it is then absorbed by living animals. The bacteria assist the process of animal digestion, whereby the vegetable molecules are altered. Upon the death of animals the current stock returns to the soil, thence to be employed by a new set of growing plants, and later by a new population of living animals.

Without microbes the "stock" would be withdrawn from circulation in living animals or vegetables, and locked up inactive in dead bodies. Upon the death of the animal, the microbes set up those processes of putrefaction and decay, whereby the stable molecules in the dead bodies become available for the food of growing plants.

CHAPTER XV.

THE ATTRACTION OF ATOMS (*continued*).

MODES OF CHEMICAL ACTION.

As a result of the operation of chemical affinity, molecules are changed in a variety of ways.

The following are some of the principal ones:—

1. Elementary or compound molecules may directly *combine*:—

Zn	+	**Cl$_2$**	=	**ZnCl$_2$**
One atom of Zinc,		One molecule of Chlorine,		One molecule of Zinc chloride,
65 parts by weight.		71 parts by weight.		136 parts by weight.
136				136

2. An element or group of elements may *displace* another element or group:—

2 HCl	+	**Zn**	=	**ZnCl$_2$**	+	**H$_2$**
Two molecules of Hydrochloric acid,		One atom of Zinc,		One molecule of Zinc chloride,		One molecule of Hydrogen,
73 parts by weight.		65 parts by weight.		136 parts by weight.		2 parts by weight.
138				138		

But in some cases the displacement may be by gradual stages. Thus marsh gas (CH$_4$) may have its hydrogen replaced by chlorine, atom by atom, until all is removed.

Thus the following compounds may be progressively formed:—

$$CH_3Cl,$$
$$CH_2Cl_2,$$
$$CHCl_3,$$
$$CCl_4.$$

3. An element or group of elements in one molecule may exchange places with an element or group of elements in another molecule:—

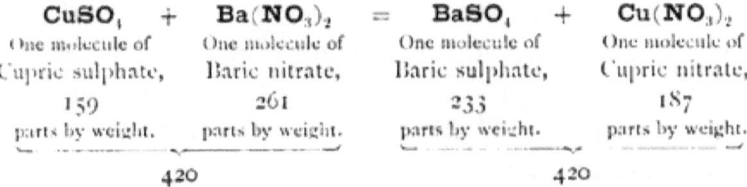

$CuSO_4$ +	$Ba(NO_3)_2$ =	$BaSO_4$ +	$Cu(NO_3)_2$
One molecule of Cupric sulphate,	One molecule of Baric nitrate,	One molecule of Baric sulphate,	One molecule of Cupric nitrate,
159 parts by weight.	261 parts by weight.	233 parts by weight.	187 parts by weight.
420		420	

4. There may be a rearrangement of elements or groups of elements within single molecules of a substance:—

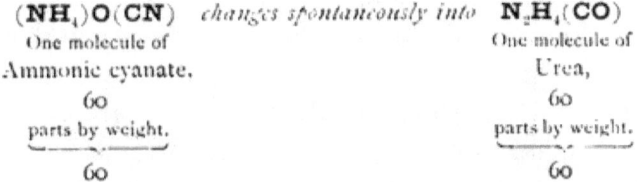

$(NH_4)O(CN)$	*changes spontaneously into*	$N_2H_4(CO)$
One molecule of Ammonic cyanate.		One molecule of Urea,
60 parts by weight.		60 parts by weight.
60		60

5. There may be a direct decomposition of a certain molecule into others of a different kind:—

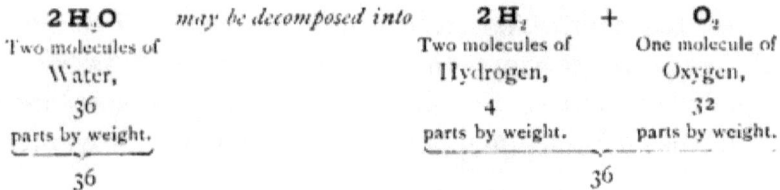

$2H_2O$	*may be decomposed into*	$2H_2$ +	O_2
Two molecules of Water,		Two molecules of Hydrogen,	One molecule of Oxygen,
36 parts by weight.		4 parts by weight.	32 parts by weight.
36		36	

The Sphere of Chemical Action. — The sphere of chemical action is evidently that of the individual molecule; as a result of chemical change, molecules change their components. This sphere is a very limited one when looked at with reference to the minuteness of a single molecule. It is one of very wide range when it is remembered that all material substances are made up of molecules, and that the character of the molecules determines the character of the mass. Chemical change, therefore, is most fundamental, altering substances in their ultimate recesses. Changing the molecules in which the identity of substances reside, it changes the identity of masses themselves. Thus all the kingdoms of nature owe to chemical action the variety of substances produced in their normal or abnormal growth, while geologic and cosmic changes involve chemical action and reaction on the largest scale.

The Results of Chemical Action. — The effects produced by chemical change are recognized as of the most striking kind; and this is true both in natural and in artificial processes.

Some of the principal effects noticed are changes of physical condition, as a substance originally solid or liquid or gaseous, at a given temperature, may change to another of these conditions; changes of color, odor, taste, or other physiologic or toxic effect; change of volume: thus sometimes chemical action draws atoms closer together. As already stated (p. 78), two volumes of hydrogen gas and one volume of oxygen gas, when chemically actuated, unite to form a new substance (water-vapor), occupying only two volumes altogether.

Sometimes there is no reduction, but rather expansion: thus gunpowder, a *solid*, experiences chemical change when slightly heated, and produces an immense volume of gas. Sometimes neither expansion nor contraction takes place.

Sometimes chemical affinity produces such violent or bizarre effects that there can be no question that it is in active exercise. In other cases, where two or more substances might be supposed to undergo chemical change, the evidences are so slight as to make the very existence of the chemical action difficult to substantiate.

Among all these various, and in many cases inexplicable results, two principles are constantly recognized, — the indestructibility of matter and the indestructibility of force.

General Laws of Chemical Action. — The following are a few general laws relating to the results of chemical action: —

The Law of Insolubility. — When there are brought together solutions that contain several elements such as would, if united, form a compound that is ordinarily insoluble in the liquid present, this insoluble compound will usually be formed and will appear as a precipitate.

This law is subject to certain limitations, yet it is of sufficiently wide application to sometimes enable the chemist to predict the formation of a given substance that may never have been produced before in that particular way.

This law finds illustration in the following equations: —

$$HCl + Ag(NO_3) = AgCl + H(NO_3);$$
$$NaCl + Ag(NO_3) = AgCl + Na(NO_3);$$
$$XCl + AgY = AgCl + XY.$$

The Law of Volatility. — When there are brought together substances whose reaction can produce a gas or a substance that is volatile at the temperature of the experiment, such volatile or gaseous substance generally will be formed, and will be liberated with effervescence.

The Indestructibility of Matter. — The amounts of weighable matter taking part in a chemical change are definite; and the sum of the weights of the products is always equal to the sum of the weights of the factors.

The Indestructibility of Energy. — The amounts of energy involved in chemical changes are definite. When the elements of a chemical compound are drawn apart, a certain amount of energy is usually absorbed. When the same elements come together to form a compound, a certain amount of energy is evolved. Now the amounts of energy in these two cases are equal.

It is true that the energy absorbed or evolved in such cases may vary in kind. It may be the energy of heat or that of light or that of electricity in some of its modifications, or it may be some combination of these. But, in any event, the facts sustain the doctrine called the conservation of energy, which involves the view that it is impossible for us to create or to destroy energy, just as it is impossible for us to create or to destroy matter. All that we can do is to change the particular form which the energy shall, for the time being, assume.

Criteria of Chemical Action. — It has already been stated as a fundamental principle that natural phenomena arrange themselves in series in which the individual members differ from their immediate neighbors by minute and sometimes almost indistinguishable details. Thus it may be expected that the drawing of distinct lines of division will often be impracticable. This statement is applicable to those different kinds of action called chemical action and physical action. While a multitude of operations are readily recognized as manifesting distinct evidences of chemical change, and at a distance from these may be produced changes that are referable distinctly to cohesive and physical forces, there are between these extremes phenomena in which the definite signs of the one or the other kind of action become less and less marked, or entirely fade away.

Thus, on a given occasion an observer may be reasonably in doubt whether certain intermingled or adjacent substances undergo or do not undergo what is properly described as chemical change.

It is therefore desirable to consider systematically the evidences upon which a decision must be reached. The following may be accepted as a guiding principle: Gain a thorough acquaintance with all the characteristics that generally attend undoubted chemical changes; then, in a doubtful case, observe whether one or several single characteristics are distinctly evident, and whether one or several characteristics can be recognized, if only in a feeble and rudimentary degree.

The following are the chief evidences of well-marked chemical action: —

(a) The generation of certain physical forces, as heat, light, electricity. These are important indications, since appliances have been produced capable of detecting very minute amounts of heat and of electrical disturbance.

(b) The production of new molecules. These may be recognized as follows: —

1. They possess new chemical composition; that is, they contain either

new elements or else they contain the original elements in new proportions by weight (and in case of gaseous elementary substances, by volume as well). This evidence involves the important law of definite proportions. (See p. 72.)

2. They possess properties differing more or less distinctly from those of the original elementary or compound molecules which, in the case in question, are supposed to have been subject to chemical change. The changes to be looked for are in the following features: —

The color; degree of opacity; refracting power for light;
The taste; physiologic and toxic properties;
The conducting power for heat, light, and electricity;
The density of the substance in the solid, liquid, or gaseous condition;
The melting and boiling points;
The degree of solubility in solvents.

CHAPTER XVI.

THE ATTRACTION OF ATOMS (*continued*).

THERMO-CHEMISTRY.

Introduction. — It has long been recognized that many chemical changes give rise to an evolution of heat. Sometimes the amount evolved is very large. Indeed, practically all man's artificial heat is the product of chemical combination.

It is also well known that the amount of heat evolved by the combustion — or other chemical reaction — of a certain weight of one substance is very different from that given out by a corresponding chemical reaction of an equal weight of another substance.

At a given moment of time any material substance or thing possesses certain internal and external relations of *parts*, and contents of heat and other *forces*, the sum total of which may be called its condition. Now any change whatever of its condition implies either some alteration of the arrangement of its *parts* externally or internally, or some alteration by increase or decrease of the amount of its *forces*. In either case the change requires for its initiation the application of some external force, which may, perhaps, be small in amount. But the change, when once it has been instituted, either absorbs or liberates a large amount of energy of some kind, generally that of heat.

Reactions of the sort referred to are sometimes classed as direct, or exothermic (those in which heat is evolved), and indirect, or endothermic (those in which heat is absorbed).

It is worthy of note that the amount of heat involved by the union of substances in an exothermic reaction is exactly equal to the amount of heat that would be required to subsequently decompose the substance produced by such operation.

The facts here stated seem to be merely forms of the general law of nature, that changes in the arrangement of material substances cannot be accomplished without the aid of force. In other words, to bring about a certain change, heat or other force must be supplied, and the energy appears to be somehow *taken in* by the compound. In the reversal of the same operation the energy taken in is given out: then force is evolved, as heat or in some other form, equivalent in amount to that originally absorbed.

Within a few years efforts have been made to learn and state with exactness the amounts of heat afforded by all the more prominent chemical actions.

The exact studies of Alexander Naumann, Julius Thomsen, Marcellin Berthelot, and others rank among the classical scientific researches of this era. The whole subject has also been carefully reviewed in recent treatises by Pattison Muir and others.

Laws of Thermo-Dynamics. — *First Law*. There is a definite quantitative relation between the amount of work done and the quantity of heat produced or destroyed.

Second Law.—If all the heat in any body or system of bodies is at the same temperature, no mechanical work can be obtained from that body or system except by bringing it into contact with another body at a lower temperature.

The important principle stated by Clerk Maxwell may with propriety be presented here:—

"The total energy of any material system is a quantity which can neither be increased nor diminished by any action between parts of the system, though it may be transformed into any of the forms of which energy is susceptible."

The Law of Maximum Work. — Berthelot states this law as follows: —

"Every chemical change accomplished without the addition of energy from without tends to the formation of that body or system of bodies, the production of which is accompanied by the evolution of the maximum quantity of heat."

Muir makes a critical examination of this statement, and thinks that Berthelot has fallen into error. Thus consider at the outset the suggestion that any chemical change can be accomplished without the addition of energy from without. Initial outward influence seems necessary to change the condition of any portions of matter not in actual process of change.

Thermal Units. — The amount of heat absorbed or evolved in chemical operations is usually represented in thermal units called *calories*.

Thomsen uses the water calory, and his unit of heat is the amount of heat necessary to raise one gramme or one kilogramme of water through one degree measured in the neighborhood of the *eighteenth to the twentieth degree* of the centigrade thermometer.

Berthelot prefers to use water at *zero degrees* centigrade.

In some cases an *ice* calory is used, the heat being measured by the amount of ice that may be changed from the solid to the liquid form without rise of temperature (one ice calory is equal to 80.025 water calories).

It should be noted, also, that sometimes the so-called large calories (C) are used, and sometimes small calories (c). The large calory relates to the kilogramme of water, the small calory to the gramme of water.

The heat of combustion of an element is sometimes defined as the amount of heat evolved by the perfect combustion of one gramme or one kilogramme of the substance.

Sometimes the heat of combustion means the quantity of heat produced by the chemical change of a number of grammes represented in a *certain reaction* of the substance.

In accordance with the first definition, the heat of combustion of hydrogen is the amount of heat produced by burning one gramme of it. In accordance with the second definition, the heat of combustion of hydrogen is the amount of heat produced by burning two grammes of it (representing one molecule of hydrogen) in accordance with the equation —

$$H_2 + O = H_2O.$$

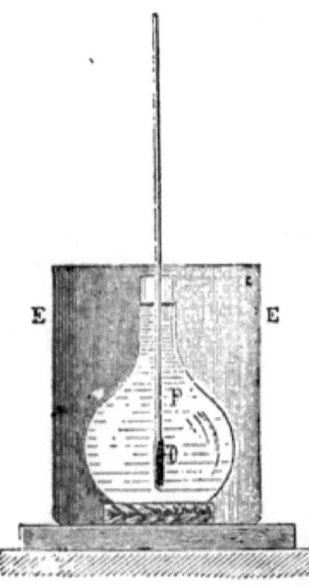

FIG. 117. — Simple form of calorimeter. The vessel *E* represents a box placed on a felt or wooden support, and provided in its interior with a non-conducting wreath for thermal insolation. The thermometer registers the heat generated in the process of a chemical reaction.

Calorimeters. — In the experiments of thermo-chemistry several different kinds of appliances have to be used.

The forms of chemical change are so varied as to the substances taking part, as to the substances ultimately produced, and as to the conditions attending the progress from the one set of substances to the other, that a thermal study of these changes demands forms of apparatus specially adapted to the different cases.

The essential parts of calorimeters are: *First*, an interior vessel of some sort, *e.g.* of glass or of platinum, in which the chemical operation proceeds. *Second*, one or more delicate and accurate thermometers, to be used in

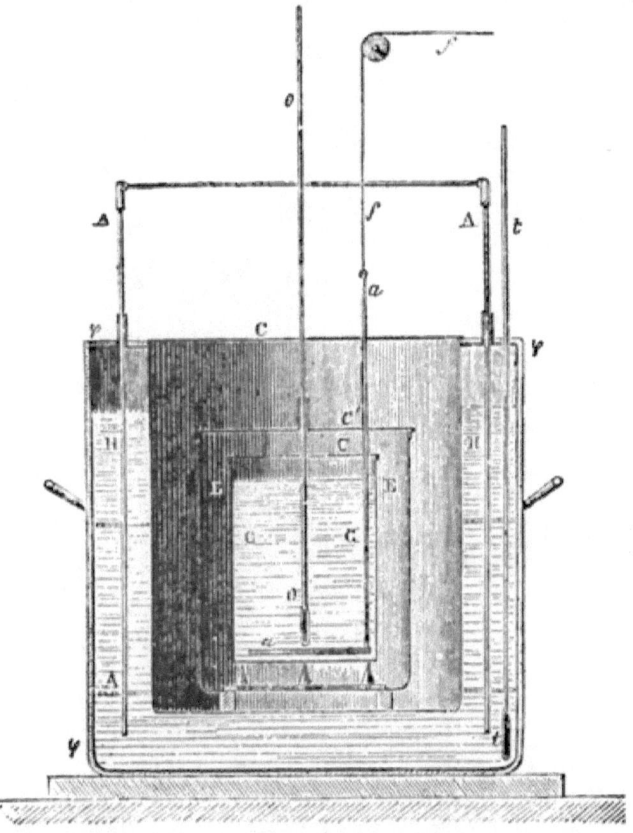

FIG. 118. — One of Berthelot's calorimeters, constructed for observing the heat generated by mixing chemical substances in solution. The outer jackets are for the purpose of preventing heat from passing from the room into the apparatus or from the apparatus outward. The beaker in which the experiment is performed is placed upon pointed supports for thermal insolation. A stirrer, *a*, is provided to diffuse throughout the solution the heat generated. A delicate thermometer suspended in the solution registers the temperature.

measuring the rise of temperature of the operation. *Third*, several protecting coatings or chambers, such

180   THE ATTRACTION OF ATOMS.

as vessels of water or of air, covered with felt. *Fourth*, mechanical stirrers to agitate the water of the outer chamber so that the heat absorbed may be evenly dis-

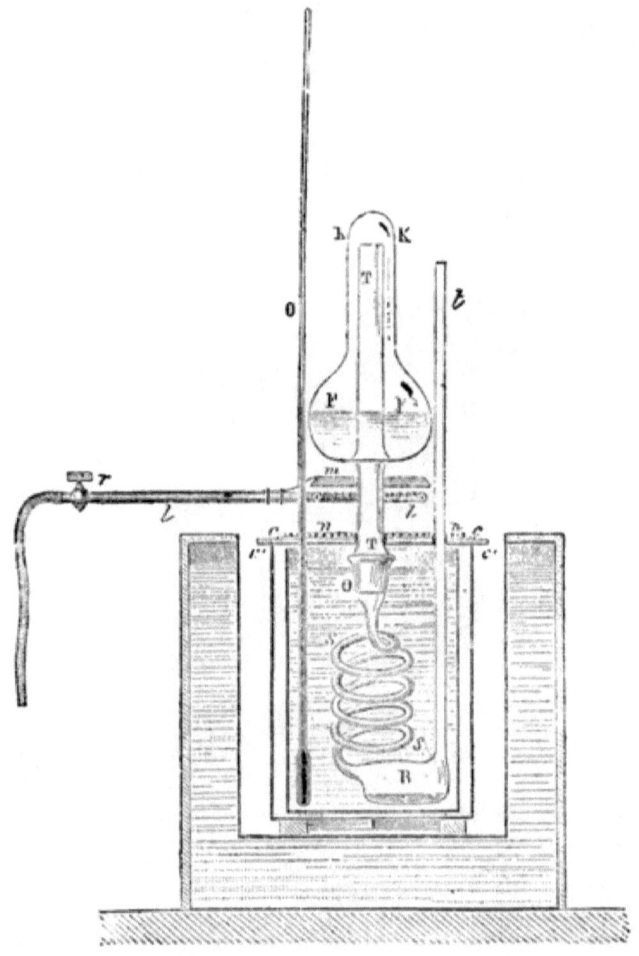

FIG. 119. — Calorimeter devised by Berthelot for experiments upon the union of certain gases. The products of chemical union are collected in the spiral S and the bulb K. The heat generated is absorbed in the liquid surrounding the spiral. The thermometer registers the heat produced. The vessel containing the spiral is thermally isolated by the several jackets placed about it.

tributed. *Fifth,* in some cases the room in which the experiments are performed is most carefully maintained at a uniform temperature. *Sixth,* sometimes the experiments have a duration of several days, so that the water employed as a protective coating (and the air of the room also) may acquire a uniform temperature.

I. One general form of apparatus is suited to experiments on the relations of solids with water or other liquids. Under this head come such cases as the solution of many salts in water, the reaction between water solutions of acids and water solutions of alkalies, and the like.

II. Another kind of apparatus is necessary in the study of combustion processes. Such are those in which solids, liquids, or gases are burned in oxygen gas, or are made to unite — by a process analogous to oxygen combustion — with sulphur, chlorine, bromine, or other substance.

III. Another form may be necessary in the study of the violent changes involved in the action of chemical agents on certain organic matters, as, for example, of nitric acid on sugar.

IV. Some chemical operations may demand apparatus for their individual treatment, as, for instance, when two gases act on each other at ordinary temperature: the union of nitrogen dioxide and oxygen affords an illustration. Again, the action of certain substances on others proceeds slowly, and special apparatus may be needed to deal with such changes: the action of oxygen on a solution of sodic thiosulphate (sodic hyposulphite) affords an illustration.

Difficulties Experienced. — Exact determinations of the amounts of heat evolved or absorbed by chemical operations are attended with certain difficulties : —

FIRST. There are the mechanical difficulties associated with the construction and use of the apparatus required.

SECOND. In certain distinctly connected series of chemical changes heat is both absorbed and evolved. The computation is thereby complicated.

The chemical union of hydrogen and chlorine would be a simple one if it were properly expressed by the equation $H + Cl = HCl$. In this expression a simple and direct union of two atoms is described.

But the true change taking place when hydrogen and chlorine unite is believed to be expressed by the equation $H_2 + Cl_2 = 2HCl$. This expression describes something more complicated than the foregoing. It expresses at least three operations; viz. a decomposition of a molecule of hydrogen, a decomposition of a molecule of chlorine, the union of an atom of hydrogen with an atom of chlorine (this particular operation being twice repeated). Now it may be safely assumed that the decomposition of the hydrogen molecule and the decomposition of the chlorine molecule both absorb heat, while the union of the atoms of hydrogen with the atoms of chlorine evolves heat. Evidently, then, the total amount of heat observed in such an operation represents *a remainder* equal to the excess of the evolved over the absorbed heat.

$H' = H - (h' + h)$.

H' represents net observed evolution of heat.

H represents amount of heat evolved by actual chemical union of the substances.

h represents amount of heat absorbed by decomposition of one of the molecules involved.

h' represents amount of heat absorbed by decomposition of the other molecule involved.

THIRD. In many cases of chemical change the heat actually generated may be partly expended in raising the *temperature* of the solids, liquids, or gases produced. Of course any such rise of temperature as is observed in the experiment must therefore be subjected to correction because of the different specific heats of the substances present. Thus the experimenter in thermo-chemistry must make a careful study of specific heat, and the results of thermo-chemistry cannot be relied upon unless correct results in specific heat are employed.

FOURTH. In many cases the final result is complicated because the substances first produced dissolve in the water present, absorbing or evolving heat by this operation.

Range of the Subject. — Evidently the field of thermo-chemistry is a wide one.

As a method of observation it has been brought to bear upon many diverse forms of action. Thus it has been applied to the study of the following classes of subjects: —

FIRST. The amounts of *heat of formation* of certain binary compounds, as the compounds of hydrogen, chlorine, bromine, iodine, sulphur, oxygen, and other elements by various methods. These include energetic changes in which various substances, elementary or compound, unite with oxygen or other elements by a process analogous to combustion.

SECOND. The *heats afforded by neutralization* of alkalies, like soda, potash, and others by acids.

THIRD. The *heats of solution* of solids, the dilution of liquids, and hydration generally.

FOURTH. The phenomena of *dissociation*, and the so-called abnormal vapor densities.

FIFTH. Certain *allotropic* and *isomeric* substances. Thus there has been made an examination of the differences in the thermal value of the combustion of different kinds of sulphur, different kinds of phosphorus, different kinds of carbon, different kinds of silicon, as well as of different isomeric compounds of the organic series, with a view of detecting, if possible, the differences of molecular structure of these elements and compounds.

Results. — While it is not possible as yet to state many distinct laws as to the heat of chemical union, it has been noticed that certain operations of a similar chemical type involve approximately equal amounts of heat, even when the particular substances taking part are all different.

The amount of heat produced when certain substances are dissolved in water affords important information as to the condition of such substances in solution. Thus it is well known that upon mingling sulphuric anhydride (SO_3) with water, heat is produced. A critical study of this operation shows plainly that a large proportion of the heat afforded is produced by the addition of the first molecule of water, and a large proportion is also produced by the addition of the second molecule of water. Muir says that there can be little or no doubt that the various results point to the formation in aqueous solutions of sulphuric anhydride (SO_3), of one definite hydrate having the formula H_2SO_4, and not of other hydrates.

The same general results are obtained by the study of the solution of other acids.

"The difference commonly expressed in the terms water of constitution and water of crystallization is evidently, so far as may be judged from thermo-chemical data, strictly a difference of degree and not of kind."

A few examples of the numerical results may be presented here.[1]

(a) *Union of Elements.*

$$H + Cl = HCl + 22,000 \text{ cal.}$$
$$H + Br = HBr + 8,440 \text{ ``}$$
$$H + I = HI - 6,040 \text{ ``}$$
$$H_2 + O = H_2O + 68,360 \text{ ``}$$
$$H_2 + S = H_2S + 4,740 \text{ ``}$$
$$S + O_2 = SO_2 + 71,080 \text{ ``}$$

(b) *Union of Compounds with Water.*

$$H_2SO_4 + H_2O = (H_2O \cdot H_2SO_4) + 6,379 \text{ cal.}$$
$$KOH \cdot 3 H_2O + 197 H_2O = (KOH \cdot 200 H_2O) + 2,751 \text{ ``}$$

---

[1] Muir, Elements of Thermal Chemistry.

THE ATTRACTION OF ATOMS. 185

(c) *Union of Acids and Bases.*

$$2\,HCl + Na_2O = 2\,NaCl + H_2O + 27{,}480 \text{ cal.}$$
$$2\,HBr + Na_2O = 2\,NaBr + H_2O + 27{,}500 \text{ ``}$$
$$2\,HI + Na_2O = 2\,NaI + H_2O + 27{,}360 \text{ ``}$$

In this last set of equations the similarity of the number of calories evolved at once suggests similarity in the operations in question. This similarity is noticeable in many other cases among elements and compounds of the same family group.

Indeed, Thomsen classifies the acids in a general way into sets somewhat as follows: —

FIRST. Acids whose heats of neutralization are about 20,000 cal. Examples are —

Nitrous acid	HNO_2,
Hypochlorous acid	$HClO$,
Carbonic acid	H_2CO_3,
Metaboric acid	$H_2B_2O_4$.

SECOND. Acids whose heats of neutralization are about 25,000 cal. Examples are —

Chromic acid	H_2CrO_4,
Succinic acid	$C_4H_6O_5$.

THIRD. Acids whose heats of neutralization are about 27,000 cal. Examples are —

Hydrochloric acid	HCl,
Hydrobromic acid	HBr,
Hydriodic acid	HI,
Nitric acid	HNO_3,
Chloric acid	$HClO_3$,
Bromic acid	$HBrO_3$,
Iodic acid	HIO_3,
Formic acid	$HOCHO$,
Acetic acid	HOC_2H_3O.

Most of the acids examined by Thomsen belong to this group.

FOURTH. Acids whose heats of neutralization are greater than 27,000 cal. (generally from 28,000 to 32,500 cal.).

Examples are —

Hydrofluoric acid	HF,
Sulphurous acid	H_2SO_3,
Sulphuric acid	H_2SO_4,
Selenic acid	H_2SeO_4,
Metaphosphoric acid	HPO_3,
Phosphorous acid	H_3PO_3,
Oxalic acid	$H_2O_2(C_2O_2)$.

CHAPTER XVII.

THE ATTRACTION OF ATOMS (*continued*).

THEORIES OF THE NATURE OF CHEMICAL ATTRACTION.

The force — whatever may be its nature — that leads substances to undergo chemical changes is often called chemical affinity. This force is capable of overcoming a certain amount of resistance; again, a certain amount of force is necessary to undo its work. It also bears definite quantitative relations to other forces, such as heat, light, and electricity: definite amounts of chemical energy are necessary to the production of unit amounts any one of them, and definite amounts of one of them are necessary to the production of a unit of chemical energy.

A very large amount of information has been secured as to its ways of working, etc., but no entirely satisfactory explanation of its *nature* has yet been offered. Views as to the nature of chemical attraction have changed from time to time as one phase of thought or another has been dominant. They have in a marked manner reflected the spirit of the time as aroused by some great discovery.

1. **Early Views.** — The general opinion of the alchemists was that somehow or other substances of like kind or origin tend to combine.

This view is evidently inadequate — the more, in that

it is now known that the most active chemical unions are between substances that are in many ways most unlike.

2. Newton's Theory. — Newton's discovery of the universal attraction of masses was naturally and easily extended to the minute particles of matter, and chemical attraction was then held to be one form of general attraction. This view was advanced by Newton, and later was supported by Berthollet. These philosophers considered, however, that the universal tendency of bodies towards each other was somewhat modified by the minuteness of the particles between which chemical changes are capable of taking place.

3. The Theory of a Special Force. — Chemical attraction has been viewed as a unique kind of force. This theory, widely accepted during the past hundred years, is that atoms and molecules are gifted by the Creator with certain specific tendencies to unite, and that with certain fixed degrees of force. This supposes the possession by atoms of an inherent property called chemical affinity. The name applies in a general way to an unexplained energy somehow residing in the atom.

4. The Electrical Theory. — Chemical attraction has been thought to be a phase of electrical energy. Davy, Dumas, Becquerel, Ampère, Berzelius, Gmelin, and others have held some form of electrical theory of chemical action. The general notion has been that atoms and molecules are naturally or may be artificially charged with varying amounts and kinds of electricity. By reason of their condition in this respect they are mutually attracted or repelled, and with varying degrees of force, somewhat as electrified masses of matter are. This theory derives support from certain important and well-defined facts. For bodies artificially electrified often manifest thereby stronger chemical attractions;

FIG. 120. — Sir Humphry Davy, Bart. Born in Penzance, England, December 17 1778; died in Geneva, Switzerland, May 29, 1829.
"Davy, when not yet thirty-two years old, occupied, in the opinion of all those who could judge of such labors, the first rank among the chemists of this or any other age."

again, chemical action yields as a product a definite quantity of electricity.

5. **The Theory of Motion.** — Williamson's theory is that chemical attraction is a form of motion. This view

FIG. 121. — André-Marie Ampère. Born at Lyons, January 20, 1775; died at Marseilles June 10, 1836. (The portrait is from a statue erected at Lyons.)

accepts the modern idea of constant atomic and molecular movement. It suggests that in atoms of all ordinary molecules a rapid but regulated interchange is going on, so that in certain cases a given atom may be continually moving from one molecule into another. A transfer of this kind could not be easily detected among molecules of the same kind, but among molecules of different kinds it would effect just such changes as are recognized in many chemical operations. Thus two or more molecules of hydrochloric acid (HCl,HCl) might make an interchange of atoms without any easily appreciable alteration of properties of the substance. But when a molecule of argentic nitrate and a molecule of hydrochloric acid are brought into contact, the interchange might produce two new molecules possessing properties easily recognized as different from those of the original two, —

$$Ag(NO_3) + HCl = AgCl + HNO_3$$
Argentic nitrate + Hydrochloric acid = Argentic chloride + Nitric acid.

This theory, moreover, not only explains such simple operations as that just referred to; it is capable of affording an adequate reason for certain of the more obscure phenomena of chemical change.

Comment on these Theories. — Chemical attraction must be looked upon as a force having in it something of general powers and something of highly specialized ones. Thus any theory of it ought to include the notion that all substances tend to come toward each other, for apparently all chemical substances will combine — there is merely a difference in the strength of this tendency in different cases. And so if masses of matter gravitate toward each other, why not molecules and even atoms? Where shall be drawn the line at which gravitative force ceases?

Further, it must be admitted that substances are found to possess certain natural qualities — inexplicable ones. What is this but a declaration that they are gifted, at their original creation, with specific powers?

Again, the close connection of electricity with chemical force is indubitable.

The modern notion of constant movement of all forms of matter applies with peculiar appropriateness to atoms and molecules, and seems to be inseparable from any idea of so intimate a change as the chemical.

There is no impropriety in considering chemical attraction as a complex rather than a simple form of force. Certainly the rich variety of its modes of action and of its results must sustain such a view.

Thus each of the theories stated contains truths. The acute observers and thinkers who have held them could not have been entirely misled. Each theory singly is merely *inadequate*. Probably in one that is adequate there must be combined the truth included in each of those stated — and more, too, as chemistry advances.

What is wanted, then, is a compact statement, sufficiently comprehensive to embrace in harmonious union the various principles known to be involved in chemical change.

CHAPTER XVIII.

ATOMIC WEIGHT.

METHOD OF WORK AND METHOD OF DESCRIPTION.

Introduction. — Whatever theory an individual may hold with respect to the existence of *atoms*, — in the most distinct import of that word, — it cannot be reasonably questioned that substances combine chemically in accordance with certain approximately fixed proportions, and that these proportions may be expressed in tolerably exact numerical form.

As a matter of fact, chemists assign to each elementary (and compound) substance a certain representative number. Such numbers are certainly *combining numbers*. They are probably much more. For elements, they represent at least an approach to *atomic weights*, and for compounds, at least an approach to *molecular weights*. They are, in fact, compact, single expressions, of the best form now known, embracing at once in harmonious union, weight ratios, volume ratios, vapor densities of elements, vapor densities of compounds, specific heats of elements, specific heats of compounds, substitution powers of elements and compounds, and even other relations besides.

"It is true it may be questioned whether there is an absolute uniformity in the mass of every ultimate atom of one and the same chemical element. Probably atomic weights merely represent a mean value around which the actual atomic weights of the atoms vary within certain narrow limits. When, therefore, it is said, *e.g.* that the atomic weight of calcium is 40, the actual fact may well be that whilst the majority of the calcium atoms really have the atomic weight of 40, some are represented by 39.9 or 40.1, a smaller number by 39.8 or 40.2, and so on. The properties which we perceive in any element are thus the mean of a number of atoms differing among themselves *very slightly*, but still not identical."[1]

In case of numbers accepted as atomic weights, the rule in this discussion is as follows: When the number involves a fractional part, this part is so modified that quantities less than .05 are rejected, while quantities equal to or greater than .05 are counted as 1 of the next higher denomination.

Practical Importance of Atomic Weights. — The numbers adopted as atomic weights have unquestioned practical value. They serve as a basis for the calculations of the chemist in his analytical processes, and also as a foundation for the work of all the great chemical manufacturing industries of the world. Their value depends in part upon the invariability of the numerical laws of combination through all chemical mutations.

Labor expended in securing Atomic Weights. — The numbers adopted have so great an importance that chemists have devoted their highest skill and their most assiduous labor to the exact ascertainment of them. (Examples may be found in the work of Berzelius on many elements, that of Stas on many elements, especially silver, of Crookes on thallium, Mallet on aluminium and gold, Cooke on antimony, and Rayleigh, Cooke, and others on oxygen.)

The different elements afford different kinds of information: Thus the atomic weight adopted to-day for one element may be worthy of far greater confidence than that adopted for another.

---

[1] W. Crookes.

And again, while the atomic weight ultimately accepted in a given case ought to be one which harmonizes with the entire body of chemical and physical knowledge, it must be expected that there should remain a few

FIG. 122. — Jons Jakob Berzelius.   Born in East Gothland (in Sweden), August 20, 1799; died August 7, 1848.

exceptional cases incapable of immediate explanation. It is acknowledged to be a matter of no slight difficulty to fix upon the atomic weight as distinguished from some simple multiple or fraction of it: in some cases, owing to insufficient data, it is at present impossible.

In any event the work demands two distinct kinds of operations: first, the experimental part, involving numerous tests and analyses; second, a work that is even higher and more difficult; *i.e.* the reasoning part — the drawing of the proper inferences from the body of experimental facts accumulated.

Since Dalton's first attempt to determine atomic weights these constants have assumed, little by little, an increasing interest. At present the effort to secure the most exact numerical expressions for them is considered by chemists a work of the highest importance.

Methods of Determination. — In determining the atomic weight of a given element, a connected series of steps must be taken. The following is a brief outline of them: —

First Step. — *Adopt a suitable unit* for the system.

Second Step. — *Fix a basis upon which shall be selected the compounds (of the element sought)* to be studied.

Third Step. — Proceed to make *gravimetric analyses* of the selected compounds. From these discover directly a *combining number* for the element.

Fourth Step. — Make choice of an *atomic weight* for the element from the various multiples or submultiples of the *combining number* discovered. In doing this, be guided by certain facts combined and applied in accordance with definite principles.

Thus, learn —

The vapor density of the element;

The vapor density of its compounds;

The volume composition of the compounds;

The specific heat of the element;
Any other suitable data.

Fifth Step. — *Confirm the foregoing choice* as fully as possible. In doing this, employ as many chemical and physical facts as possible. With this in view, study the element and its compounds in connection with molecular formulas.

These involve a consideration of —
Molecular grouping;
Specific heats of compounds;
The boiling-points of compounds;
The crystalline forms;
Such other relationships as may be useful.

Sixth Step. — Bring all the atomic weights obtained *into one table* and arrange them in an appropriate order. This has been attempted by Newlands, Mendeléeff,[1] and many others. The two investigators mentioned have been among the most successful.

NOTE. It may be noted here that it is the gravimetric analyses that give the results that are numerically capable of the highest degree of accuracy. Indeed, they afford the only decisive foundation. The numbers they afford are certainly *combining numbers*.

On the other hand, the various vapor densities and the various specific heats are mainly valuable as guiding in the choice between the several multiples of a number already learned.

The Method of Discussion. — It is plain that the subject in hand is an extended one. There is some difficulty even in the selection of a method of presenting it. Several ways are open.

It is *not* well to follow here the exact historical course of the subject, for this has been marked by tentative and even erroneous views. However instructive these may be to the experienced chemist, they can only embarrass the beginner.

---

[1] Transliteration of Russian words: Nature, xli. 396; xlii. 6, 77, 316.

It seems preferable to carry the student over a course directly leading to what is now accepted as truth; but the course to be selected should accord with the natural progress and be shaped by a sound pedagogic method.

In the following discussion certain numbers are very quickly adopted.

This is not improper. But it should not be forgotten that, as already intimated, the historical progress of chemistry has involved a much more circuitous route to reach these numbers than the discussion suggests.

CHAPTER XIX.

ATOMIC WEIGHT (*continued*).

FIRST STEP: A UNIT ADOPTED.

The subject involves the search for certain numbers. Now it must be remembered that all numerical expressions of mixed mathematics involve the use — either expressed or implied — of some unit; and often the unit chosen depends more upon convenience than upon any other consideration. These statements apply to atomic weights. Some unit has to be selected. At the present day the one almost universally accepted is the weight of a single atom of hydrogen, — a weight extremely small, but one that it is possible (though not necessary) to express in terms of every-day weights.

It is believed that the weight of a single atom of hydrogen is equal to 35 grammes divided by 10^{23}, — an amount about equal to one and one quarter ounces divided by one million million million million. This minute weight has received the special name *microcrith*. When, therefore, it is said that the atomic weight of oxygen is 16, the meaning is that a single atom of oxygen weighs 16 microcriths.

A Different Unit might be used. — It must be remembered that this selection of the atom of hydrogen as the standard is a matter of convenience partly, being based on the fact that no other atom has so low a weight. The weight of any other element might be used if it were found to be more convenient.

Dalton used hydrogen in his first table of atomic weights. Subsequently Berzelius and others recommended using oxygen as the standard, calling its weight 100.[1] This latter system, however, was found to have the objection of affording in many cases numbers too large for convenience. Thus, when the atomic weight of oxygen equals 100, the atomic weight of uranium becomes about 1494.

Certain practical objections have been urged recently against the employment of the atom of hydrogen as the unit with the atomic weight, 1.

Thus it is very difficult to decide upon the exact ratio of the weight of the hydrogen atom to that of the oxygen atom. A number of ratios have been obtained as a result of extremely careful investigation. Thus it has been placed as low as $H:O::1:15.869$, and as high as $H:O::1:16.010$. But since oxygen is the starting-point for the determination of the atomic weights of a great many other elements, any error in the adopted ratio of $H:O$ is transferred to nearly all the other atomic weights. Thus, if $O = 15.869$, then the atomic weight of uranium $= 237.14$. If $O = 16.010$, the atomic weight of uranium $= 239.25$.

To avoid this difficulty it has been proposed, while using hydrogen as the nominal basis of the system, to use oxygen as the practical basis; to call the atomic weight of oxygen 16, and then to hereafter determine what the exact atomic weight of hydrogen is. At present it would be about 1.0025; but it might be expected to be slightly modified from time to time as the ratio of the combining portions of oxygen and hydrogen is determined with greater and greater exactness. Meanwhile, changes in this ratio would not necessarily alter the entire scheme of atomic weights.

This proposition, to adopt oxygen with atomic weight 16 as the unit, has received the approval of many eminent chemists.

SECOND STEP: SELECTION OF THE COMPOUNDS AND THE PROCESSES TO BE EMPLOYED.

In a certain sense every chemical compound (and every chemical process) is capable of contributing some-

---

[1] Meyer, L., and Seubert, K., American Chemical Journal, vii. 96.

thing to the knowledge of atomic weights; and, in fact, the study of a large number of them is necessary. But certain ones are far more serviceable than others.

The compounds must be selected, then, with definite principles in mind, and they should conform to as many as possible of the following requisites: —

1. They should be such as can be prepared in a form possessing a high degree of purity.

Moreover, in purifying the materials used for analysis the so-called "fractional" methods should be employed.

2. As many different compounds as practicable should be tested. This helps the analyst to avoid "constant" sources of error.

3. They should be such as are capable of having their composition determined with a high degree of exactness.

With this in view the analyses should require as simple processes as possible.

"Improvements made of late in manipulative methods and apparatus have tended to reduce very much the magnitude of what are commonly called 'fortuitous' errors in quantitative determinations of matter, and to increase greatly the accuracy. No one nowadays would undertake the determination of an atomic weight of one of the better-known elements without taking such elaborate precautions as must practically insure pretty close concordance of results when obtained by the same method applied in the same hands. But such mere close agreement is not alone sufficient."

4. They should be such as possess a molecular condition that is simple and can be distinctly ascertained.

Thus it is well to employ compounds *of only two elements*. Again, if a given pair of elements forms a series of compounds, that one containing the smallest amount of the element under consideration is most likely to contain one atom of it.

Again, compounds that are *gaseous* at ordinary temperatures or that can be easily vaporized, are employed. (See p. 37.)

5. In reactions depended upon, only such other elements should be concerned as may be counted among those of which the atomic weights are already known with the nearest approach to exactness.

Thus, (*a*) Compounds with hydrogen are preferred when they are practicable; for hydrogen has generally been adopted as the basis of the system, and so involves little or no error.

Indeed, there are four compounds of hydrogen so well suited to this purpose that they have been called the type compounds of modern chemistry; they are —

Hydrochloric acid (hydrogen and chlorine);
Water (hydrogen and oxygen);
Ammonia gas (hydrogen and nitrogen);
Marsh gas (hydrogen and carbon).

(*b*) In case hydrogen compounds are impracticable, then the compounds selected should, if possible, be such as contain certain atoms that have been compared directly and quantitatively with hydrogen.

The majority of the elements do not form hydrogen compounds, so that, in fact, recourse is oftener had to oxygen compounds and chlorine compounds. All the elements (except fluorine, and that combines with hydrogen) form oxides, and oxygen itself has been compared with hydrogen with a high degree of accuracy. Again, chlorine combines with a great many metals, and the atomic weight of chlorine has been accurately determined.

If oxygen is adopted (with the number 16) as the basis of the system, then compounds of oxygen will be selected, even in preference to compounds of hydrogen.

6. Further, it is desirable that as few other elements as possible — the assumed atomic weights of which will have to be taken into account — shall be involved in each *single* reaction depended upon.

7. In selecting *different* processes to be applied to the determination of the atomic weight of a given element, it is desirable that not the same, but as many

different other elements as possible, shall be concerned in the several reactions.

Study of the Reactions. — Careful preliminary study is required as to the general effect of each reaction involved, and as to how it may be influenced by the conditions of the experiment. For it has been learned more and more of late that many reactions — perhaps it should rather be said all reactions — which have been generally supposed to be of the simplest nature, are, in reality, complex.

As many different and independent processes as can be devised (reasonably free from apparent sources of error) should be employed.

Each process employed should be as simple as possible, both in the *kind of chemical changes* involved as well as in liability to manipulative errors.

Comparison of Results. — In comparison of results, careful consideration should be given as to the probable influence of each kind of experiment; *i.e.* whether it tends on the whole to yield higher results or lower results than the truth.[1]

---

[1] Mallet, J. W., American Chemical Journal, xii. 82.

CHAPTER XX.

ATOMIC WEIGHT (*continued*).

THIRD STEP: THE EXPERIMENTAL WORK FOR SECURING A FEW ATOMIC WEIGHTS.

This stage is a very extended one. It involves, in fact, all the experimental work done and recorded — up to the time of forming conclusions. Now chemical work has been done — varying in extent and accuracy — upon nearly if not all substances known to civilized nations. But of course, for the purpose of this discussion, only a few of the results can be referred to.

I.

A STUDY OF CHLORINE AND ITS AFFILIATED ELEMENTS, BROMINE AND IODINE (ALSO SODIUM, POTASSIUM, AND SILVER).

(*a*) **Gravimetric Composition of Certain Compounds.** — *Experimental Facts.* A study of certain compounds of chlorine, bromine, and iodine has afforded a series of facts which are stated in the following table : —

FIRST TABLE.

Approximate Percentage Composition of Twelve Compounds. — Experimental Results.

Chloride of Hydrogen.		Bromide of Hydrogen.		Iodide of Hydrogen.	
Hydrogen,	2.75 p. ct.	Hydrogen,	1.24 p. ct.	Hydrogen,	.79 p. ct.
Chlorine,	97.25 "	Bromine,	98.76 "	Iodine,	99.21 "
	100.00		100.00		100.00

Chloride of Sodium.		Bromide of Sodium.		Iodide of Sodium.	
Sodium,	39.38 p. ct.	Sodium,	22.37 p. ct.	Sodium,	15.37 p. ct
Chlorine,	60.62 "	Bromine,	77.63 "	Iodine,	84.63 "
	100.00		100.00		100.00

Chloride of Potassium.		Bromide of Potassium.		Iodide of Potassium.	
Potassium,	52.42 p. ct.	Potassium,	32.83 p. ct.	Potassium,	23.55 p. ct.
Chlorine,	47.58 "	Bromine,	67.17 "	Iodine,	76.45 "
	100.00		100.00		100.00

Chloride of Silver.		Bromide of Silver.		Iodide of Silver.	
Silver,	75.26 p. ct.	Silver,	57.44 p. ct.	Silver,	45.97 p. ct.
Chlorine,	24.74 "	Bromine,	42.56 "	Iodine,	54.03 "
	100.00		100.00		100.00

Experimental Results of the First Table stated differently. — A consideration of the direct results given in Table 1 leads to the detection of the following facts: —

1. The numbers fall into series.

2. In each of the series of chlorides, bromides, and iodides, the chlorine is smaller in amount than the bromine, and the bromine is smaller in amount than the iodine.

3. When the amounts of hydrogen, sodium, potassium, and silver are compared, it is seen that their quantities are in the order stated; hydrogen being in smallest amount, and silver in largest.

4. If the three hydrogen compounds are compared on the basis of one part of hydrogen, the hydrogen series of compounds shows the following composition : —

SECOND TABLE.

Hydrogen Compounds. — Experimental Results.

Hydrochloric Acid.		*Hydrobromic Acid.*		*Hydriodic Acid.*	
Hydrogen,	1.	Hydrogen,	1.	Hydrogen,	1.
Chlorine,	35.4	Bromine,	79.8	Iodine,	126.6
	36.4		80.8		127.6

5. If trial is made with the numbers obtained in Table 2 (viz. 35.4 for chlorine, 79.8 for bromine, and 126.6 for iodine) in the other compounds under consideration, the following very remarkable results are obtained:—

THIRD TABLE.

Sodium Compounds. — Experimental Results.

Sodic Chloride.		*Sodic Bromide.*		*Sodic Iodide.*	
Sodium,	23.	Sodium,	23.	Sodium,	23.
Chlorine,	35.4	Bromine,	79.8	Iodine,	126.6
	58.4		102.8		149.6

FOURTH TABLE.

Potassium Compounds. — Experimental Results.

Potassic Chloride.		*Potassic Bromide.*		*Potassic Iodide.*	
Potassium,	39.	Potassium,	39.	Potassium,	39.
Chlorine,	35.4	Bromine,	79.8	Iodine,	126.6
	74.4		118.8		165.6

FIFTH TABLE.

Silver Compounds. — Experimental Results.

Argentic Chloride.		*Argentic Bromide.*		*Argentic Iodide.*	
Silver,	107.7	Silver,	107.7	Silver,	107.7
Chlorine,	35.4	Bromine,	79.8	Iodine,	126.6
	143.1		187.5		234.3

The results are brought together in the following table :—

SIXTH TABLE.

Experimental Results.

CHLORIDES.			BROMIDES.			IODIDES.		
Hydric Chloride.			*Hydric Bromide.*			*Hydric Iodide.*		
	Per cent.	Ratio.		Per cent.	Ratio.		Per cent.	Ratio.
Hydrogen,	2.75	1.	Hydrogen,	1.24	1.	Hydrogen,	.79	1.
Chlorine,	97.25	35.4	Bromine,	98.76	79.8	Iodine,	99.21	126.6
	100.00	36.4		100.00	80.8		100.00	127.6
Sodic Chloride.			*Sodic Bromide.*			*Sodic Iodide.*		
	Per cent.	Ratio.		Per cent.	Ratio.		Per cent.	Ratio.
Sodium,	39.38	23.	Sodium,	22.37	23.	Sodium,	15.37	23.
Chlorine,	60.62	35.4	Bromine,	77.63	79.8	Iodine,	84.63	126.6
	100.00	58.4		100.00	102.8		100.00	149.6
Potassic Chloride.			*Potassic Bromide.*			*Potassic Iodide.*		
	Per cent.	Ratio.		Per cent.	Ratio.		Per cent.	Ratio.
Potassium,	52.42	39.	Potassium,	32.83	39.	Potassium,	23.55	39.
Chlorine,	47.58	35.4	Bromine,	67.17	79.8	Iodine,	76.45	126.6
	100.00	74.4		100.00	118.8		100.00	165.6
Argentic Chloride.			*Argentic Bromide.*			*Argentic Iodide.*		
	Per cent.	Ratio.		Per cent.	Ratio.		Per cent.	Ratio.
Silver,	75.26	107.7	Silver,	57.44	107.7	Silver,	45.97	107.7
Chlorine,	24.74	35.4	Bromine,	42.56	79.8	Iodine,	54.03	126.6
	100.00	143.1		100.00	187.5		100.00	234.3

Inference 1. — Evidently, then, the following numbers have some important fundamental meaning: —

NUMBERS WORTHY OF CONSIDERATION.

Hydrogen, H, adopted as . . . . 1.

{ Chlorine, Cl, found to be, ... 35.4
  Bromine, Br,   "   "   " 79.8
  Iodine, I,   "   "   " 126.6

{ Sodium, Na,   "   "   "   . . . . . 23.
  Potassium, K, "   "   "   . . . . . 39.
  Silver, Ag,   "   "   " 107.7

(It may be noted here, as a fact, that subsequent study and comparison of all results accessible confirm the opinion that these numbers are important, and are probably atomic weights.)

Inference 2. — These results give the following as molecular weights: —

Hydrochloric acid 36.4
Sodic chloride 58.4
Potassic chloride 74.4
Argentic chloride 143.1

Hydrobromic acid 80.8
Sodic bromide 102.8
Potassic bromide 118.8
Argentic bromide 187.5

Hydriodic acid 127.6
Sodic iodide 149.6
Potassic iodide 165.6
Argentic iodide 234.3

Inference 3. — These results suggest the following formulas: —

H Cl	H Br	H I
Na Cl	Na Br	Na I
K Cl	K Br	K I
Ag Cl	Ag Br	Ag I

CHAPTER XXI.

ATOMIC WEIGHT (*continued*).

FOURTH STEP: THE CHOICE OF A PARTICULAR ATOMIC WEIGHT FROM SEVERAL COMBINING NUMBERS.

(*b*) **The Density of Certain Elementary Gases and Vapors.** — *Experimental Fact* 1. When chlorine gas is weighed it is found to weigh, volume for volume, about 35.4 times as much as hydrogen. Hydrogen is usually taken as a standard of comparison for weight of gases.

The number 35.4 is called the *density* of chlorine gas.

Experimental Fact 2. — A similar experiment is tried with bromine vapor. Its vapor density is found to be about 79.9.

Experimental Fact 3. — A similar experiment is tried with iodine vapor. Its vapor density is found to be about 127.

Inference 1. — Certain numbers are obtained that are worthy of attention. Evidently their similarity to the numbers already obtained are not mere coincidences.

Inference 2. — The density of an elementary gaseous substance at once gives its atomic weight.

Inference 3. — The gaseous state is a very favorable one for study in this connection: in this state bodies appear to be in a sort of equality of condition that favors their comparison with each other from the point, perhaps, of even other relations than atomic weight merely.

(*c*) **The Volume Composition.** — *Experimental Fact.* When hydrochloric acid gas is tested, it is found that two volumes of the gas yield by decomposition one volume of hydrogen gas and one volume of chlorine gas.

Inference 1. — This sustains the view assumed in the preceding study, that the compound called hydrochloric acid consists of one atom of hydrogen and one atom of chlorine. (Of course it is possible that the volume

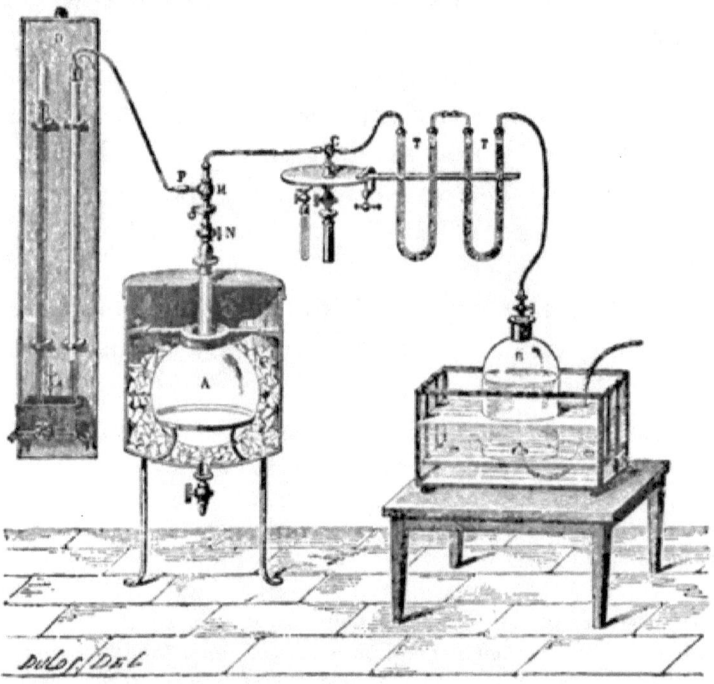

Fig. 123. — Regnault's apparatus for filling a globe with gas at the temperature of 0° C. previous to weighing, for the purpose of determining a vapor density.

composition merely teaches that the number of atoms of hydrogen and of chlorine are equal; that the formula is HCl or H_2Cl_2 or H_3Cl_3 or H_nCl_n. Until further information is secured, chemists assume the truth of the simplest expression. Further information is not in fact wanting, for subsequent study strongly sustains the view that the formula is indeed HCl, and no other. See p. 228.)

Experimental Facts. — Similar results for hydrobromic acid and hydriodic acid serve to confirm the figures already accepted for bromine and iodine.

Inference 2.—These facts sustain all the inferences previously reached.

FIG. 124.—Henri Victor Regnault. Distinguished French physicist. Born at Aix-la-Chapelle in 1810; died in 1878.

(*d*) **The Vapor Density of Certain Compound Substances.** —*Experimental Facts.* The vapor density of hydrochloric acid gas is found experimentally to be about 18. This means that a given volume of hydrochloric acid gas weighs 18 times as much as the same volume of hydrogen gas.

Evidently there exists a very simple relation between the number representing the density of hydrochloric acid gas and the number already adopted as its molecular weight; *i.e.*: —

      18   :   36   ::   1   :   2

  Density of   Molecular weight of
hydrochloric acid gas. hydrochloric acid gas.

FIG. 125. — Method of introducing into the bulb *A* a portion of liquid, *C*, whose vapor density is to be subsequently determined by Dumas' method.

Inference 1. — Perhaps in case of other *compound substances* the vapor density and molecular weight are connected by the same simple relation. Perhaps, as a rule, the molecular weight (which is, from its nature, difficult to determine) may be obtained by multiplying by 2 the vapor density (which in many cases it is easy to determine). (By subsequent experiment this principle appears to be sustained, and the inference is accepted as a just one.)

Inference 2. — Perhaps, in case of *elementary substances*, the vapor density and molecular weight are in the ratio above suggested; that is 1 : 2. (By subsequent experiment this principle appears to be well grounded, and the inference is accepted as a just one.) A remarkable result follows.

The vapor density of chlorine gas is found experimentally to be 35.4; then the molecular weight is $35.4 \times 2 = 70.8$. If, then, the molecular weight is 70.8 (and the atomic weight is 35.4), then the number of atoms in the molecules is 2, and the molecular formula of chlorine is Cl_2.

(This generalization is a very important one. By subsequent experiment it appears to be sustained with respect to most of the elements capable of existing in the form of gas or vapor. See p. 249.)

(*c*) **The Specific Heats of Elements.** — *Definition.* The specific heat of a substance is the amount of heat necessary to raise one unit of weight of the substance one degree of temperature. Water has a high specific heat, and as it is usually taken as the standard, the specific heats of most other substances are expressed by decimal fractions. (See p. 46.)

Experimental Facts. — Dulong and Petit made a great many experimental determinations of the specific heats of solid substances. (It is more difficult with liquids and gases.) In case of elementary solids, whose atomic

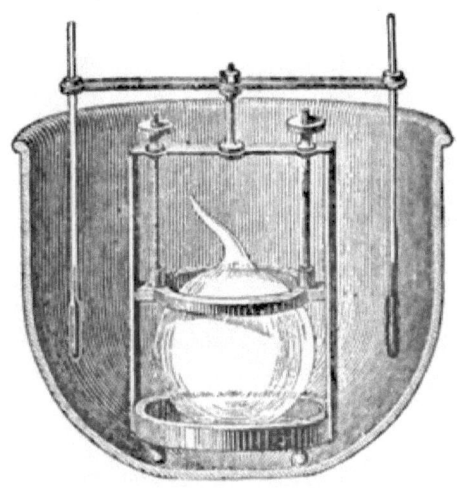

FIG. 126. — Hot bath, provided with thermometers for determination of vapor densities by Dumas' method. A small but weighed portion of liquid is placed in the globe. It is then vaporized by the heat of the bath. Subsequently the tip of the flask is sealed, whereupon the weight of vapor present at a certain observed temperature may be determined by the balance.

weights had been previously determined *by other methods*, they found that apparently the higher the atomic weight, the lower the specific heat.

Inference. — They then enunciated the following law, now called the law of Dulong and Petit: —

The specific heats of solid elements are inversely proportional to their atomic weights.

The law is likewise expressed in the following proportion: —

Specific heat of A : Specific heat of B : : Atomic weight of B : Atomic weight of A.

This proportion also discloses the following fact: The product of the specific heat of any solid element by its atomic weight is a constant number.

This constant is found to be about 6.3.

Again, the constant 6.3 divided by the specific heat of a solid element yields as a quotient the atomic weight of the element.

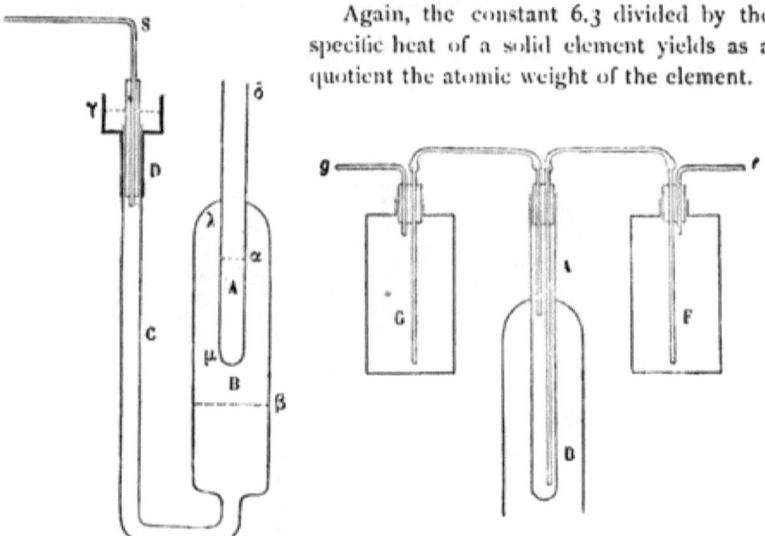

FIG. 127. — Bunsen's ice calorimeter for determining specific heats of substances.

FIG. 128. — Cooling apparatus for Bunsen's calorimeter, already referred to, Fig. 127.

Bunsen's method of determining specific heats of substances may be explained by reference to Figs. 127 and 128. S is a glass tube carefully graduated or calibrated. D is an iron reservoir containing mercury; the latter extends from the line β through the tube C up into the tube S. The space B is filled with water. A is at first empty. In using the apparatus, intensely cold alcohol is passed in a current through the tube A by use of the apparatus, Fig. 128. In due time the water in B is frozen completely. The cold alcohol is then withdrawn. Next the entire apparatus is placed in melting snow or ice to bring the whole to zero centigrade. Next the position of the summit of the mercury column in the tube S is carefully observed. Now one gramme of water at 100° C. is placed in the tube A. The water melts a certain portion of the ice in *B*. Thereupon contraction takes place, as is indicated by the rise of mercury in *B* and the fall of mercury in *S*. The point to which the summit of the mercury column in *S* now retracts is observed. The distance this summit has traversed corresponds to the specific heat of water. Subsequently the substance to be tested is raised to the temperature of 100° C. Then one gramme of it is placed in the tube *A*. Thereupon it melts another portion of ice. Further retraction of mercury takes place. The amount of such retraction being measured and compared with the retraction due to one gramme of water, gives directly the specific heat of the substance in question.

Specific Heats of Compounds. — The specific heats of many *solid compounds* have been determined. In a few cases they are in general harmony with the law of Dulong and Petit; in many cases they are not. In some cases the specific heat of a solid compound containing two atoms, and having a given molecular weight, appears to be twice as great as that of a single elementary substance of the same atomic weight; and the specific heat of a solid compound of three atoms appears to be three times as great as that of a mere element whose atomic weight equals the molecular weight of the compound. This seems to show that when a compound body is heated, the rise of temperature is associated with a motion of each different atom in the molecule, and not merely with that of the molecule as a whole.

While, then, it requires a certain amount of heat to impart to an elementary substance an amount of motion sufficient to produce a certain change of temperature, in case of a compound body with the same molecular weight more heat is required to impart to it the amount of motion that affords the same temperature as that already supposed, — twice as much heat is required for compounds of two atoms, and three times as much heat for compounds of three atoms.

Atomic Heats of Elements. — The constant 6.3 is often called the atomic heat of an element. The significance of this expression may be explained as follows: Upon taking one unit of weight of each of several elementary substances, and then applying equal amounts of heat to each of them, it is observed that the temperature rises to different degrees in the different cases.

Suppose, however, different weights of the substances are experimented upon, say —

> 7 parts of lithium,
> 56 parts of iron,
> 194 parts of platinum,
> 204 parts of lead;

it is then found that equal amounts of heat added to these several amounts by weight produce in all the same advances of temperature. Evidently, then, equal amounts of heat applied to single atoms of these substances will produce the same advances of temperature. This explains the statement that the atomic heats of the elements are the same.

Special Cases. — *Indirect Determination of Specific Heats.* In cases of certain elements the specific heat cannot readily be determined *directly*. This is especially true of the *gaseous* elements, as hydrogen, fluorine, chlorine, oxygen, nitrogen. But indirect methods have been devised. (In case of certain solid elements, as carbon, boron, and silicon, the specific heats are abnormal. This is supposed to be due to the tendency of these elements to assume allotropic modifications.)

Indirect Method for Chlorine. — The specific heat of argentic chloride has been learned experimentally. It is .089. Now by independent methods the molecular weight of the compound is found to be 143.1, and it is found to consist of two atoms, — one of silver and one of chlorine. Multiplying the molecular weight by the specific heat ($143.1 \times .089$), the molecular heat 12.7 is obtained. Subtracting from this number the atomic heat of silver, 6.1 (as experimentally obtained), there remains 6.6 as the atomic heat of chlorine indirectly determined.

Many other similar indirect determinations for chlorine have been made; generally speaking, they yield the number 6.4.

Indirect Method for Carbon. — The specific heat of carbon hexachloride (C_2Cl_6) has been learned experimentally. It is .177. By independent methods, its molecular weight is found to be 236.4, and it is found to consist of eight atoms as stated. Multiplying the molecular weight by the specific heat ($236.4 \times .177$), the molecular heat, 41.8, is obtained. Subtracting from this number six times the atomic heat of chlorine ($6 \times 6.4 =$

38.4) (41.8 − 38.4 = 3.4), there remains 3.4, which is twice the atomic heat of carbon. By this means the atomic heat of carbon is 1.7 indirectly determined.

Other similar indirect determinations have given the atomic heat of carbon, in combination, as about 2.

Specific Heat of Bromine. — *Experimental Fact.* The specific heat of solid bromine has been found by direct experiment; it is .08432. Applying the law of Dulong and Petit, *i.e.* dividing .08432 into the constant 6.3, and there results the quotient 75.

Inference. — The atomic weight of bromine is nearly 75.

But the studies of bromine already referred to (pp. 204 and 209) show that the combining number is about 79.8. Its atomic weight is probably either 79.8, or 2×79.8, or 3×79.8, or $n \times 79.8$. Now, as has been said before, while the specific heat does not give the exact atomic weight, it enables us to decide that 79.8, and not some multiple (or fraction) of it, should be accepted.

Specific Heat of Iodine. — *Experimental Fact.* The specific heat of solid iodine has been found by direct experiment to be .0541. Applying the law of Dulong and Petit, $\frac{6.3}{.0541}$ = about 116.

Inference. — The atomic weight of iodine is about 116.

But the results previously and otherwise obtained point to 126.6. Evidently the specific heat confirms this selection.

Specific Heat of Sodium. — *Experimental Fact.* The specific heat of sodium is found directly to be .293. Now $\frac{6.3}{.293}$ = about 22.

Inference. — The atomic weight of sodium is about 22.

But results previously and otherwise obtained have given the number 23. Evidently the specific heat confirms this selection.

Specific Heat of Potassium. — *Experimental Fact.* The specific heat of potassium is found directly to be .166. Now $\frac{6.3}{.166}$ = about 38.

Inference. — The atomic weight of potassium is about 38.
But results previously and otherwise obtained have given the number 39. Evidently the specific heat confirms this selection.

Specific Heat of Silver. — *Experimental Fact.* The specific heat of silver is found directly to be .057. Now $\frac{6.3}{.057}$ about 111.

Inference. — The atomic weight of silver is about 111.
But results previously and otherwise obtained have given the number 107.7. Evidently the specific heat confirms this selection.

General Inference. — If now reference is made to the provisional table presented on p. 208, it is seen that the numbers there given secure marked confirmation from the experiments subsequently detailed.

They are repeated here as —

SEVENTH TABLE.

WELL-ESTABLISHED ATOMIC WEIGHTS.

Atomic weight of hydrogen adopted as					. . . .	1.
"	"	" chlorine found to be		. . . .	35.4
"	"	" bromine	"		. .	79.8
"	"	" iodine	"		. . . .	126.6
"	"	" sodium	"		. . . .	23.0
"	"	" potassium	"		. . . .	39.0
"	"	" silver	"		. .	107.7

II.

A STUDY OF OXYGEN AND SOME OF ITS COMPOUNDS.

The study of chlorine and its affiliated elements has afforded a considerable number of suggestions. These may well be applied to other elements. Oxygen may well be studied first, because of its great importance in this as well as other relations.

(*a*) **Experimental Fact.** — The density of oxygen gas is about 16.

Inference. — The atomic weight of oxygen is about 16.

(*b*) **Experimental Fact.** — The volume composition of water vapor is as follows: two volumes of hydrogen and one volume of oxygen.

Inference 1. — The formula of water is H_2O (see pp. 226 and 229).
Inference 2. — The molecular weight of water is about $2 + 16 = 18$.

(*c*) **Experimental Fact.** — The vapor density of water vapor is about 9.

Inference. — The molecular weight of water is about $9 \times 2 =$ about 18.
This inference is based on results obtained under chlorine. It sustains the views already adopted in this section.

(*d*) **Experimental Fact.** — Gravimetric analysis shows that water is made up as follows: —

Hydrogen	. .	11.11 parts by weight.
Oxygen	.	88.88 " "
		99.99

The ratios are as 1 : 8 or 2 : 16.

Inference. — These facts add support to the views previously accepted, and contribute greatly to create confidence in the general principles as well as the numerical results adopted.

(*c*) **Experimental Facts**. — Oxygen forms compounds with sodium, potassium, and silver, having the composition given in the following table. (As a matter of fact it forms many others, but this set is selected as affording a strict continuity to the argument.)

EIGHTH TABLE.
Percentage Basis.

Sodium and Oxygen.		Potassium and Oxygen.		Silver and Oxygen.	
	Per cents.		Per cents.		Per cents.
Sodium,	74.19	Potassium,	82.98	Silver,	93.09
Oxygen,	25.81	Oxygen,	17.02	Oxygen,	6.91
	100.00		100.00		100.00

If now the results of the eighth table are computed on another basis, *i.e.*, using the atomic weights accepted for sodium 23, for potassium 39, and for silver 107.7, there is afforded a new table. Its results are surprising, but they present a strict statement of *facts* — in a special form merely.

NINTH TABLE.
Atomic Weight Basis.

Sodium and Oxygen.		Potassium and Oxygen.		Silver and Oxygen.	
Sodium,	23	Potassium,	39	Silver,	107.7
Oxygen,	8	Oxygen,	8	Oxygen,	8.
	31		47		115.7

Inference. — Either the atomic weight of oxygen is 8 instead of 16; or, if it is indeed 16, then the atomic weights accepted for sodium, potassium, and silver are only one-half what they should be; *or else the compounds have the formulas* Na_2O, K_2O, Ag_2O, respectively. This latter supposition satisfies all the foregoing facts (and many others) so well, that it has been universally adopted, and with it the atomic weight 16 (or thereabouts) for oxygen.

On this view the following table may be arranged. It is, as to *numerical relations*, a strict statement of experimental facts.

TENTH TABLE.

NEW FORMULAS.

Na_2, $23 \times 2 = 46$ K_2, $39 \times 2 = 78$ Ag_2, $107.7 \times 2 = 215.4$
O,   16   O,   16   O,   16.
              ——   ——   ——
              62   94   231.4

Inferences from the Tenth Table. — From various facts already presented, the following two groups of formulas have been accepted : —

Hydrochloric acid,	HCl	Hydrogen oxide (water),	H_2O
Hydrobromic acid,	HBr		
Hydriodic acid,	HI		
Sodic chloride,	NaCl	Sodic oxide,	Na_2O
Sodic bromide,	NaBr		
Sodic iodide,	NaI		
Potassic chloride,	KCl	Potassic oxide,	K_2O
Potassic bromide,	KBr		
Potassic iodide,	KI		
Argentic chloride,	AgCl	Argentic oxide,	Ag_2O
Argentic bromide,	AgBr		
Argentic iodide,	AgI		

An inspection of these formulas recalls the striking and very important suggestion that oxygen possesses *a different numerical nature from chlorine, bromine, and iodine*, not only in that it has a different atomic weight from theirs, but, further, in this, that while chlorine, bromine, and iodine are satisfied to combine with hydrogen, sodium, potassium, and silver, atom for atom, the atom of oxygen is only satisfied when it combines with two atoms of the elements in question.

This fact is emphasized by a great many other compounds, — so much so that oxygen is called a *dyad*, and the other elements mentioned are called *monads*. Further, an atom of oxygen is said to have two points of attraction, while an atom of hydrogen (and one atom of each of the other substances mentioned) is said to have one point of attraction.

Yet further, the property of the elementary atoms by virtue of which they attract different numbers of atoms is called *equivalence* or *valence*. And an atom of hydrogen is accepted as the unit of valence. (See p. 223.)

III.

A STUDY OF SULPHUR AND SOME OF ITS COMPOUNDS.

Sulphur may be discussed very much as oxygen has been.

(*a*) **Experimental Fact.** — The density of sulphur vapor is found to be about 32.2.

Inference. — The atomic weight of sulphur is about 32.

(*b*) **Experimental Fact.** — Sulphuretted hydrogen gas is found to be composed of hydrogen and sulphur, and to have the density about 17.2.

Inference 1. — Its molecular weight is 34.
Inference 2. — It is probably made up of one atom of sulphur weighing 32, and two atoms of hydrogen weighing 2.
Inference 3. — Its formula is probably H_2S.

(*c*) **Experimental Fact.** — Sulphuretted hydrogen gas has been found to have the percentage composition: —

Hydrogen . . .	5.88	parts by weight.
Sulphur	94.12	" "
Total	100.	

The ratios of these numbers are evidently as 1 : 16, or as 2 : 32.

Inference. — These facts sustain the views already presented as to the composition of sulphuretted hydrogen, and that the atomic weight of sulphur is 32.

(*d*) **Experimental Fact.** — The specific heat of sulphur has been found to be .188. But $\dfrac{6.3}{.188}$ = about 33.

Inference. — This affirms the selection of the number 32 as the atomic weight of sulphur.

Experimental Fact. — The most exact determinations of the atomic weight of sulphur have been based upon its combination with silver. The composition of sulphide of silver has been learned by experiment to be —

Silver	87.07 per cent.
Sulphur	12.93 "
	100.

Inference 1. — If the atomic weight previously found for silver is 107.7, then the atomic weight of sulphur is 16, or some multiple of it. But the results already stated suggest the number 32 as the proper atomic weight. If this view is accepted, the following inference may be obtained.

Inference 2. — The formula of sulphide of silver is Ag_2S. Then —

The molecular weight of Ag_2 = 215.4 (87.07 per cent)
The atomic weight of S = 32. (12.93 per cent)
$\overline{}$
247.4 (100. per cent)

Inference 3. — From these results sulphur is placed in the class of dyads, as oxygen was. The notion of valence already reached is therefore sustained by the studies of sulphur above described.

Inference 4. — As previously intimated, it has been learned that certain elements — hydrogen, chlorine, bromine, iodine, sodium, potassium, silver — have the equivalence *one*, and certain elements — oxygen, sulphur — have the equivalence *two*. The suggestion naturally arises that perhaps other elements have the equivalence *three* or *four*, or indeed higher numbers yet. It may be added that this suggestion is amply sustained by facts, some of which will be presented soon. The general notion of valence is adopted as a fundamental fact of chemistry.

NOTE. By gravimetric analysis of binary compounds containing on the one hand elements whose atomic weights have been provisionally adopted as just described, and on the other hand other elements, atomic weights of these latter elements may be secured — subject, of course, to revision in the light of additional facts such as have been already presented.

CHAPTER XXII.

ATOMIC WEIGHT (*continued*).

FIFTH STEP: CONFIRMATION OF THE ATOMIC WEIGHTS CHOSEN.

The preceding chapters on atomic weights have sufficed to show how a few of these important constants can be secured. The discussion has called attention to certain methods pursued and certain precepts accepted.

Of course, when the atomic weight of a substance is determined by a sufficient number of independent methods, such weight may be used in fixing the molecular formula of compounds of the element.

On the other hand, it is a fact of deeper significance that *when the molecular formula of a compound* can be determined by independent methods, this formula may be of great service in securing atomic weights of elements, or in confirming those already secured. Every effort is made, therefore, to determine molecular formulas of elements and compounds.

I.

MOLECULAR FORMULA SECURED BY VOLUME COMPOSITION.

In illustration, the four type-compounds of modern chemistry may be referred to —

Hydrochloric Acid. — *First Fact.* Hydrochloric acid consists of hydrogen and chlorine, and nothing else.

Second Fact. — Hydrochloric acid is a gas.

Third Fact. — Two volumes of hydrochloric acid gas yield by decomposition one volume of hydrogen and one volume of chlorine.

Inference. — The molecule of hydrochloric acid has some one of the following formulas: —

$$HCl,$$
$$H_2Cl_2,$$
$$H_3Cl_3,$$
$$\ldots\ldots,$$
$$H_nCl_n.$$

On a previous page certain reasons have been assigned for adopting the formula HCl. Yet other reasons will be assigned hereafter. (See p. 228.)

Fourth Fact. — Hydrogen and chlorine do not form any compound but this one, called hydrochloric acid.

Inference. — Hydrochloric acid is the simplest possible compound of the elements hydrogen and chlorine. Therefore it is a compound containing one atom of each.

NOTE. This inference is not a particularly valid one, but it naturally arises in cases of a single compound of two elementary substances.

In some cases it is distinctly misleading. Thus, previous to the recognition of hydrogen dioxide, but one compound of hydrogen and oxygen was known; viz. water. On this general ground the formula HO was adopted. Subsequently the volume relations and other considerations (some of them stated in this chapter) have led to the adoption of the formula H_2O.

Water. — *First Fact.* Water consists of hydrogen and oxygen, and nothing else.

Second Fact. — Water may be changed into a vapor and experimented upon in that form.

Third Fact. — Two volumes of water vapor yield, by

decomposition, two volumes of hydrogen gas and one volume of oxygen gas.

Inference. — The formula of water is probably one of the following: —

$$H_2O,$$
$$H_4O_2,$$
$$H_6O_3,$$
$$\ldots\ldots,$$
$$H_{2n}O_n.$$

But reasons have been heretofore stated, favoring the view that the formula is H_2O. Hereafter other reasons will be stated for this view.

Ammonia Gas. — *First Fact.* Ammonia gas consists of hydrogen and nitrogen, and nothing else.

Second Fact. — The substance is a gas.

Third Fact. — When two volumes of ammonia gas are decomposed, they afford one volume of nitrogen gas and three volumes of hydrogen gas.

Inference. — The formula for ammonia gas is one of the following: —

$$H_3N,$$
$$H_6N_2,$$
$$H_9N_3,$$
$$\ldots\ldots,$$
$$H_{3n}N_n.$$

The simplest formula, H_3N, may be accepted for the present, with the intention of changing it if facts hereafter discovered demand such change.

Marsh Gas. — *First Fact.* Marsh gas consists of carbon and hydrogen, and nothing else.

Second Fact. — The substance is a gas.

Third Fact. — Two volumes of marsh gas yield, by decomposition, four volumes of hydrogen gas. (The volume relations of the carbon cannot be stated, since carbon cannot be obtained in a state of gas.)

Inference. — The formula of marsh gas is one of the following: —

$$H_4C,$$
$$H_8C_2,$$
$$H_{12}C_3,$$
$$\ldots\ldots,$$
$$H_{4n}C_n.$$

The simplest formula H_4C may be adopted at present, with the intention of changing it if facts subsequently discovered demand such change.

(Evidently the fact that carbon is not obtainable in a gaseous form diminishes, to some extent, confidence in the formula adopted.)

II.

MOLECULAR FORMULA BASED UPON CHEMICAL SUBSTITUTION.

Hydrochloric Acid. — *First Fact.* Many chlorides may be formed by substituting certain metals for the hydrogen in hydrochloric acid: such are the well-known chlorides, sodium chloride, potassium chloride, silver chloride.

Second Fact. — When such substitutions as those just referred to are made, it is found that *the whole* of the hydrogen may be replaced by a metal, but *no fractional part* can be. Thus it is not possible to form a chloride in which part of the hydrogen has been replaced by potassium and part of the hydrogen remains unreplaced. Replacement must be of the *whole of the hydrogen or of none at all.*

Inference 1. — The amount of hydrogen in hydrochloric acid is chemically indivisible. In other words, it is an atom.

NOTE. It is true that the chemist cannot so carry out his experiment as to work upon a single molecule of hydrochloric acid. The smallest quantity upon which he can experiment must necessarily contain millions of molecules. But this does not in any way invalidate the conclusions just reached.

If in a mass of hydrochloric acid containing (say) four molecules of hydrochloric acid, one-half of the hydrogen were replaced by potassium, there might be a change somewhat as indicated below: —

First Stage.	Second Stage.
HCl,	HCl,
HCl,	HCl,
HCl,	KCl,
HCl.	KCl.

In a sense, however, all the hydrogen in the hydrochloric acid *experimented upon* has been replaced by potassium. But the formulas given represent simply an incomplete operation. They do not represent *a new and single* compound containing part hydrogen and part potassium. Instead, they represent a *mixture* of compounds, the one having all its hydrogen replaced by potassium, as already intimated, the other having none of its hydrogen yet replaced.

By a process of experiment and reasoning entirely similar to the foregoing, it may be shown that the *chlorine* in hydrochloric acid may be replaced by another element; for example, bromine or iodine. It is found, however, that bromine and iodine respectively replace the whole of the chlorine in the hydrochloric acid, or none at all. They cannot replace the one-half or one-fourth, nor any other fractional part of the chlorine.

Inference 2. — The chlorine in the molecule of hydrochloric acid is practically indivisible. That is, it is a single atom.

Thus it seems to be proved that the formula for hydrochloric acid is HCl.

Water. — *First Fact.* Many compounds may be formed by substituting proper metals for the hydrogen in water. Thus the well-known compounds, potassic oxide, sodic oxide, may be easily formed.

Second Fact. — When such substitutions as those just referred to are made, it is found that they may be accomplished in two different ways ; *i.e.* either the whole of the hydrogen may be replaced by the potassium and sodium, respectively, or the half of the hydrogen may be so replaced. In these cases two entirely different substances are produced. When the whole of the hydrogen is replaced, the substance called potassic oxide is produced. It contains 78 parts by weight of potassium, and 16 parts by weight of oxygen. When the half of hydrogen is replaced, a distinct and well-known compound is produced, called potassium hydroxide. It contains 39 parts by weight of potassium, 1 part by weight of hydrogen, 16 parts by weight of oxygen.

Here, then, it is seen that potassium may replace the half of the hydrogen in water or the whole of it. But the displacement of hydrogen in water is not possible *in any other fractional way.*

By a course of experiment and reasoning similar to that pursued with hydrochloric acid, it may be shown that the oxygen of water is not divisible; in other words, is a single atom.

Inference 1. — It appears that the one-half of the hydrogen that is in a molecule of water is the indivisible portion of hydrogen; *i.e.* is one atom. In other words, it appears that the molecule of water has two atoms of hydrogen.

Inference 2. — It appears that the proper formula for water is H_2O.

Ammonia Gas. — By the use of undoubted facts of observation and a method of reasoning similar to that pursued in the two preceding illustrations, it may be shown that the hydrogen of ammonia gas is divisible into three parts, and that its nitrogen is not divisible.

Thus the inference is secured that the formula for ammonia gas is H_3N.

Marsh Gas. — In similar fashion, marsh gas may be shown to have an amount of hydrogen that is divisible into four parts and not into any other fractional amounts, and that its carbon is indivisible. (See p. 168.)

Thus it is apparent that the formula H_4C should be adopted for marsh gas.

The foregoing discussion has sustained an opinion previously expressed with reference to the valence of chlorine and oxygen; viz. chlorine has been made out to be a monad, oxygen a dyad, and so also nitrogen and carbon respectively a triad and a tetrad.

Many other facts of chemical substitution could be presented to sustain the views just enunciated, and thence, of course, to sustain the formulas already adopted.

III.

RELATION OF MOLECULAR WEIGHT TO MELTING-POINTS OF SOLIDS AND BOILING-POINTS OF LIQUIDS.

Chemistry as a system cannot be complete until elements and compounds can be arranged in orderly lists, showing regular progression of the various physical and chemical properties of the substances.

Many small groups are recognized where such a principle as this is successfully carried out.

There are certain marked cases of the simultaneous advance of molecular weights and boiling-points in the case of elementary substances. For example note the following:—

Name.	Atomic Weight.	Molecular Weight.	Boiling-point.
Chlorine,	35.4	70.8	−33.60° C.
Bromine,	79.8	159.6	59.27° C.
Iodine,	126.6	253.2	250° C.

Such advance, however, is by no means uniform. If the list of *elementary substances* be arranged in an order commencing with that of lowest atomic weight, and ending with that of highest atomic weight, it will be found that while there is a general tendency toward increase of melting-point (for most of the elements are solid at ordinary temperature), this increase is by no means uniform or even regular.

The facts with respect to *compound substances* are so marked, however, that the chances are that the order of arrangement of elements by atomic weights is not quite correct; that some of the elements in the solid state may have a larger number of atoms than others; and so the proper molecular weight of elements (at present in most cases unknown) may be at present recognized only in those few that can be given in regular order of melting-points.

Here is a group showing such progress: —

Name.	Formula.	Molecular Weight.	Boiling-point.
Sulphur dioxide,	SO_2	64	$-10°$ C.
Sulphur trioxide,	SO_3	80	$46°$ C.

In case, however, of the compounds of carbon and hydrogen, several series can be constructed which advance with very striking regularity at once in molecular weight and in boiling-point.

Name.	Formula.	Molecular Weight.	Boiling-point.
Methane (marsh gas),	CH_4	16	(gas)
Ethane,	C_2H_6	30	(gas)
Propane,	C_3H_8	44	(gas)
Butane,	C_4H_{10}	58	$1°$ C.
Pentane,	C_5H_{12}	72	$38°$ C.
Hexane,	C_6H_{14}	86	$71.5°$ C.
Heptane,	C_7H_{16}	100	$98.4°$ C.
Octane,	C_8H_{18}	114	$125.5°$ C.

A Study of Certain Nitrogen Compounds. — One use that can be made of boiling-point may be illustrated by a short study of the compounds of nitrogen and oxygen.

These compounds are as follows: —

Nitrogen monoxide, N_2O; vapor density, 21.99; melting-point . $-99°$ C.
Nitrogen dioxide, N_2O_2 or NO, not liquefied at . . . . . . $-110°$ C.
Nitrogen trioxide, N_2O_3; vapor density, 37.95; liquefies . . . $-10°$ C.

Nitrogen tetroxide, N_2O_4 or NO_2; v. d. anomalous $\begin{cases} \text{solidifies at} & -9° C. \\ \text{liquid boils at} & 22° C. \end{cases}$

Nitrogen pentoxide, N_2O_5; solid melts at . . . . 30° C.

The vapor density for nitrogen monoxide (about 22) points to a molecular weight 44, and this corresponds to the requirements of the formula N_2O. The vapor density for nitrogen trioxide (about 38) points to the molecular weight 76. This corresponds to the requirements of the formula N_2O_3.

There are reasons that need not be specified here for adopting for nitrogen pentoxide the formula N_2O_5.

The following question then arises with respect to the two compounds remaining:—

Is the formula N_2O_2 or the formula NO to be preferred for nitrogen dioxide? Comparing the boiling-points of this compound with the compound N_2O, called nitrogen protoxide, it is seen that the boiling-point of N_2O is very much lower. In accordance with a general rule, substances having lower boiling-points should be of simpler constitution. A substance having the formula N_2O_2 is of more complex constitution than a substance having the formula N_2O. If, however, we assign to the substance designated as N_2O_2 the formula NO, its constitution becomes simpler than that of N_2O; it then accords with the general rule.

There are certain facts with respect to the substance designated as N_2O_4 that lead to the conclusion that this is the correct formula at *low temperatures*. It is thought at *high temperatures* it dissociates, forming the simpler molecule NO_2. (See p. 144.)

NOTE. It appears probable that the following general law may be safely accepted:—

General Law. — The more complex compounds of a series condense more easily to liquids and solids and decompose more readily by heat than the less complex compounds.

IV.

RELATION OF MOLECULAR FORMULA TO CRYSTALLINE FORM.

Mitscherlich's law of isomorphism may be stated as follows: —

Fig. 129. — Mass of alum crystals.

In general, when two solid compounds are isomorphous — that is, have the same crystalline form, — they have the same number of atoms and the same molecular arrangement.

It hardly needs mention that in applying the law it must be remembered that compound radicles, like ammonium, methylamine, ethylamine, etc., must be counted as elements.

It must be admitted that there are well-recognized cases of substances of similar crystalline form that are evidently not chemically analogous; again, substances possessing most marked chemical resemblances are known that solidify in differing crystalline forms. But while there are many exceptions to the law, and it is not an authoritative guide, yet it is of occasional value in confirming results obtained by other methods.

1. The law is well illustrated by the alums. A certain substance called alum has been recognized with more or less distinctness for at least two thousand years. During the last century its chemical composition has been distinctly made out. It is usually expressed by the formula

$$K_2SO_4 \cdot Al_2(SO_4)_3 \cdot 24\,H_2O.$$

FIG. 130. — Diagrams showing different forms assumed by crystals of the first or regular system.

It crystallizes easily and distinctly in cubes or regular octohedrons, or some simple modification of these belonging to the first or regular system.

Within a few years at least a dozen substances have been recognized which bear such marked structural resemblance to ordinary alum that they have all been called *alums*.

Examples: —

Potassio-aluminic alum (ordinary alum), $K_2SO_4 \cdot Al_2(SO_4)_3 \cdot 24\,H_2O.$
Sodio-aluminic alum, $Na_2SO_4 \cdot Al_2(SO_4)_3 \cdot 24\,H_2O.$
Ammonio-aluminic alum, $(NH_4)_2SO_4 \cdot Al_2(SO_4)_3 \cdot 24\,H_2O.$
Potassio-chromic alum, $K_2SO_4 \cdot Cr_2(SO_4)_3 \cdot 24\,H_2O.$
Ammonio-ferric alum. $(NH_4)_2SO_4 \cdot Fe_2(SO_4)_3 \cdot 24\,H_2O.$

The fact that these all crystallize in form similar to ordinary alum leads chemists to confidently adopt for them the same general molecular formula as that assigned to ordinary alum. Yet, further, two other inferences are drawn; viz. that potassium, sodium, and ammonium are radicles of analogous character, and that aluminium, chromium, and iron are also of analogous chemical character.

2. The substances calcic carbonate, potassic nitrate, and sodic nitrate have some marked crystalline resemblances. Thus a certain form of calcic carbonate, found crystallized in nature and called *arragonite*, is recognized as having the same crystalline form as potassic nitrate. Again, a slightly different form of calcic carbonate, though of the same chemical composition, but found crystallized in nature in the form called *calcspar*, has a crystalline form similar to that of common sodic nitrate.

Here, then, are three substances closely related as to crystalline forms. Analysis has shown that possible formulas for these substances are —

$$CaCO_3,$$
$$KNO_3,$$
$$NaNO_3.$$

The crystalline resemblances then favor the adoption of these formulas. It may be added that the formula $CaCO_3$ leads to the approval of the number 40 as the atomic weight of calcium, and the other formulas favor the employment of the numbers 23 and 39, already well substantiated, for sodium and potassium respectively.

V.

RELATION OF MOLECULAR FORMULA TO MOLECULAR STABILITY.

The question has arisen whether the substance known as nitrogen dioxide should have the formula N_2O_2, or the formula NO. Facts already given, under the head of *boiling-points*, indicate pretty clearly that the formula should be NO. Certain facts with respect to the decomposition of this substance as compared with the decomposition of the substance nitrogen monoxide (N_2O) sustain this view.

Thus the general rule is that *substances of more complex composition possess less chemical stability* than substances of less complex composition. Now when a piece of glowing phosphorus is plunged into the gas nitrogen monoxide (N_2O), it readily decomposes the molecule, withdrawing oxygen;

and the phosphorus continues to burn by combining with this oxygen. But when burning phosphorus is introduced into the gas called nitrogen dioxide (N_2O_2 or NO), it is extinguished; that is, it does not withdraw the oxygen. The conclusion is that the molecule of nitrogen dioxide is of greater chemical stability than the molecule N_2O, and therefore is of simpler constitution than the molecule N_2O. This favors the assumption that the formula of nitrogen dioxide should be taken as NO, and not as N_2O_2. For if the formula were accepted as N_2O_2, we should have the more complex molecule, having the greater chemical stability instead of the less, as the general law demands. (See p. 232.)

VI.

MOLECULAR FORMULAS SUGGESTED BY RELATIONSHIP.

The substances marsh gas, methyl alcohol, and formic acid are naturally related. The accepted formulas are as follows: —

Marsh gas . . . CH_4,
Methyl alcohol . CH_3OH,
Formic acid . . . $HOCHO$, or $HCOOH$.

So the substances ethylene, ethyl alcohol, and acetic acid are naturally related. The accepted formulas for these are as follows: —

Ethylene . . . C_2H_4,
Ethyl alcohol . . C_2H_5OH,
Acetic acid . . . $HO(C_2H_3O)$, or CH_3COOH.

Now the formulas chiefly in question are those of formic acid and acetic acid. But the fact that the marsh gas and methyl alcohol, whose formulas are established, have each one atom of carbon, favors the assumption that formic acid contains one atom of carbon, and therefore that it has the formula assigned.

Again, the fact that ethylene and ethyl alcohol, whose formulas are well established, have each two atoms of carbon, favors the assumption that acetic acid has two atoms of carbon, and therefore that it has the formula assigned.

VII.

MOLECULAR FORMULA SUGGESTED BY PRODUCTS OF DECOMPOSITION.

When the two organic compounds, formic acid and acetic acid, are made to combine with alkaline substances to produce respectively formates and acetates, it is observed that 46 parts of formic acid do the same work as 60 parts of acetic acid. These numbers may then be taken temporarily as the molecular weights of the two substances. Taken in conjunction with other facts of analysis, the following statement may be prepared: —

Formic Acid. — The formula HCOOH corresponds to molecular weight 46.

Acetic Acid. — The formula CH_3COOH corresponds to molecular weight 60.

But analysis has shown that 46 parts of formic acid contain 12 parts of carbon, and that 60 parts of acetic acid contain 24 parts of carbon.

Next consider the products of decomposition.

Experiment has shown that when these acids are subjected to the current of the galvanic battery they are decomposed. Now it is observed that from formic acid but one carbon compound is produced, *i.e.* carbon dioxide. But from acetic acid two compounds of carbon are produced, — carbon monoxide and ethane. These facts suggest that the carbon in formic acid acts somehow as a unit, while in acetic acid there is such a difference in the condition of the carbon that it is easily susceptible of division into at least two parts, the one part doing one thing, the other part doing another thing.

These facts — while not absolutely conclusive — favor the opinion that formic acid contains one atom of carbon, while acetic acid contains two atoms of carbon.

VIII.

THE ADOPTED MOLECULAR FORMULA SUPPORTED BY CERTAIN EXCEPTIONAL COMPOUNDS.

Certain double salts, produced naturally or artificially, may point out the existence of definite molecular groups.

Thus in the Solvay soda works, at Syracuse, N.Y., a salt was artificially formed in considerable quantity (although decomposable by water) which had practically the following formula: $MgCO_3 \cdot Na_2CO_3 \cdot NaCl$. This combination was evidently of those single groups which are accepted as molecules. Thus it seems to establish the three formulas adopted for the three substances taking part in it.[1]

IX.

OTHER ILLUSTRATIONS OF THE CLOSE CONNECTION BETWEEN THE PROPERTIES OF SUBSTANCES AND THEIR MOLECULAR WEIGHTS.

(*a*) **Density of Liquids as related to their Molecular Structure.** — The following law, called Groshans's law, has been enunciated: —

At the temperature of ebullition, the density of compound bodies in the liquid state is in proportion to the number of the atoms in their molecules.

(*b*) **The Relation of the Physiological Action of Inorganic Compounds with their Molecular Weights.** — In a study of the influence of different salts in solution when introduced into the blood of living animals, it has been observed that the acid radicle of the salt has but little

---

[1] Chemical News, lvii. 3.

influence. Any physiological action produced depends almost entirely on the electro-positive component of the salt, *i.e.* upon the metal. Again, it has been noted that practically all the elements found in organized bodies have atomic weights of less than 40. Thus it appears that all the positive elements among them are monads and dyads. Now, it has been noted that the physiological action of substances increases from the monads onward.

Dr. J. Blake, who has studied this subject, points out that the monads tend to affect but one set of tissues or organs, — the pulmonary arteries. The dyads affect two or more, — *i.e.* the centres of vomiting, the voluntary and cardiac muscles, — while with elements of higher equivalence the influence is more widespread and therefore more considerable: it extends to the ganglia and even the brain itself. Again, experiments seem to show that the physiological efficiency of substances belonging to one and the same isomorphous group is directly proportional to the atomic weights; *i.e.* the higher the atomic weight, the greater the action. The law has been studied with respect to the following substances: *first*, lithium, sodium, rubidium, thallium, silver; *second*, magnesium, iron, manganese, cobalt, nickel, copper, zinc, cadmium; *third*, calcium, strontium, barium, lead; *fourth*, palladium, platinum, osmium, gold.

In case of chlorine, bromine, and iodine, however, the action seems to vary inversely as the atomic weights.

In case of potassium and ammonium the influence is also partially exceptional.

The whole study is an important one and may be looked upon as likely to offer more valuable information in the future. Thus it may be that the biological relations of chemical substances may assist in determining the position of atoms and molecules in the chemical scale.[1]

(*c*) **The Magnetic Rotary Polarization of Compounds as related to their Chemical Constitution.** — Chemists have long felt assured that such rotation was dependent

---

[1] Blake, J., Chemical News, xliii. 191; xlv. 111; lvii. 194.

upon the kind of molecules involved as well as their number, yet difficulties in the way of demonstration have prevented the attainment of any satisfactory conclusions.

Dr. Perkin has studied this subject very carefully. He has adopted a new system of unit lengths for those portions of the substances experimented upon; that is, he has employed such portions of liquid compounds *as would produce unit lengths of columns of vapor* when in the latter condition. As a result, it appears that certain definite relations do exist between this magnetic rotary power and the molecular constitution of bodies. It is not practicable to express these results in a few words. Reference, therefore, must be made to Perkin's original paper.[1]

(*d*) **Freezing-points of Solutions as related to the Molecular Weights of the Substances dissolved: Raoult's Method.** — This method is of chief importance in case of compounds which cannot be vaporized without dissociation. It is based upon a principle described by Coppet, that "salts of analogous constitution, dissolved in quantities proportional to their molecular weights, produce in their solution the same depression of the freezing-point of the solvent."

Raoult has made a careful study of this subject, and as a result of his work the law seems now to meet general acceptance.

If C represents the depression in degrees centigrade, P the number of grammes of substance dissolved in 100 grammes of water, and M the molecular weight of the substance, —

$\frac{C}{P} = A$, the coefficient of depression of the substance;

$MA = T$, the molecular depression of the solvent.

The constant T varies according to the substances used, and according to the solvents employed. Of the latter, water, benzene, and acetic acid have been found most useful.

---

[1] Perkin, W. H., Journal of the London Chemical Society, 1884, Transactions, p. 421.

This method is of especial value in its application to many organic compounds of high molecular weight.

One general result of Raoult's studies may be presented in the following general form: —

While the freezing-point of a pure liquid is constant, every molecule of foreign matter that dissolves occasions the same constant depression of that point. In other words, in dilute solutions the depression of the freezing-point of the solvent varies with the ratio between the numbers of molecules of solvent and numbers of molecules of substance dissolved.

As respects vapor tension of the liquid, Raoult has shown that its depression bears a relation to the percentage of foreign molecules which is independent of temperature, but if certain proper distinctions are made, is such that it represents a relative depression which may become a definite constant for any substance which that particular liquid may dissolve.

Raoult's method has been applied to the determination of the molecular weights of certain of the carbohydrates. As a result, the molecular weight of dextrin was found to correspond approximately to the number 6480. This points to the formula 20 $(C_{12}H_{20}O_{10})$. In similar fashion the formula for soluble starch has been suggested to be five times this value, or 5 $(C_{12}H_{20}O_{10})_{20}$. This would give as a molecular weight of soluble starch about the number 32,400.[1]

---

[1] American Chemical Journal, xi. 67; also xii. 130 and 142. Chemical News, lx. 66.

CHAPTER XXIII.

ATOMIC WEIGHT (continued).

SIXTH STEP: BRING ALL THE ATOMIC WEIGHTS INTO ONE TABULAR STATEMENT. THE PERIODIC LAW.

WHEN the atomic weights of practically all the elements have been provisionally adopted, by methods involving the principles already referred to (and perhaps some others), then the "periodic law" may be considered. By its use certain numbers already adopted may be confirmed; perhaps, on the other hand, it may determine a new selection from the various possible atomic weights.

The Work of Newlands and of Mendeléeff. — The English chemist Newlands (in 1863–64) and the Russian chemist Mendeléeff (in 1869–70) independently published tables of the elements known at about those dates. They first arranged all known elements in a long list, commencing with that of the lowest atomic weight, and advancing numerically to that of the highest. In considering this list, they noticed that the elements formed several natural series, the members of a given series showing a periodic progress in chemical properties; for example, in the kind and amount of atom-fixing power. Upon arranging the several series one above another, it at once appeared that the corresponding members of the several series formed natural groups. For example, calcium, strontium, and barium appeared in one group; so also did phosphorus, arsenic, and antimony; and also sulphur, selenium, and tellurium; and also chlorine, bromine, and iodine.

Professor Crookes remarks: "Undoubtedly one of the grandest steps taken in pure chemistry within our epoch has been the discovery of the *periodic law*. This generalization (as reference to *Chemical News* will show, vols. vii., x., xii., and xiii.) was in the first place due to Mr. J. A. R. Newlands. It was some time afterwards independently discovered by Mendeléeff, and since been developed both by that eminent savant and by Meyer and Carnelley."

The total result of the Mendeléeff classification is now known as "the periodic system." It is presented in the table (including 68 elements) found on the next page.

As soon as the periodic table was adjusted, it suggested the important truth that "the properties of an element are a periodic function of its atomic weight." It now appears, therefore, that when a new element is studied and its properties are learned, these properties determine its place in the periodic table. But the place in the table at once suggests which, among several multiples, shall be accepted as the atomic weight.

The system has now secured very general adoption. Some of its merits are the following: —

FIRST. It is based on the atomic weights, — constants which it must be assumed are dependent upon some fundamental characteristics of elements.

PERIODIC TABLE OF ATOMIC WEIGHTS.

SERIES.	I. R₂O	II. R₂O₂ or RO	III. R₂O₃	IV. RH₄, R₂O₄ or RO₂	V. RH₃, R₂O₅	VI. RH₂, R₂O₆ or RO₃	VII. RH, R₂O₇	VIII. (R₂H) hyd. com., R₂O₈ higher or RO₄ oxy. com.
1	1 H	—	—	—	—	—	—	—
2	Li 7	Be 9	B 11	C 12	N 14	O 16	F 19	—
3	Na 23	Mg 24	Al 27	Si 28	P 31	S 32	Cl 35·4	—
4	K 39	Ca 40	Sc 44	Ti 48	V 51	Cr 52	Mn 54	Fe 56, Ni 58, Co 59
5	Cu 63	Zn 65	Ga 69	Ge 72	As 75	Se 78	Br 80	—
6	Rb 85	Sr 87	Y 89	Zr 90	Nb 94	Mo 96	—	Ru 104, Rh 104, Pd 106
7	Ag 108	Cd 112	In 114	Sn 118	Sb 120	Te 125	I 127	—
8	Cs 133	Ba 137	La 138	Ce 140	Di 145	—	Sm 150	—
9	—	—	Yb 173	—	Ta 182	W 184	—	—
10	—	—	—	—	—	—	—	—
11	Au 196	Hg 200	Tl 204	Pb 207	Bi 208	—	—	Os 199, Ir 193, Pt 194
12	—	—	—	Th 233	—	U 239	—	—

Second. It throws most of the elements into groups and series which accord with many of their undoubted geological, physical, and chemical properties.

It cannot be denied that in this system some elements are brought together that do not appear to be closely related. This is merely equivalent to admitting that the system is not yet perfect. Probably, also, some so-called elements are wrongly placed because of their peculiar compound nature.

Third. It helps in the decision as to which of several combining numbers of an element shall be accepted as its atomic weight. Thus indium might have the number 75.6 or the number 113.4 (one and a-half times the former). The latter number is now chosen under guidance of the periodic law.

Fourth. The periodic table has shown some gaps in the series of numbers representing atomic weights. On this basis — as long ago as 1871 — Mendeléeff predicted the existence of two new elements, and more, he stated their general range of properties. To one he gave the provisional name eka-aluminium. Now in 1876 the element gallium was discovered, and it proved to be the predicted eka-aluminium. So scandium, discovered in 1879, proved to be Mendeléeff's predicted eka-boron.

The recently discovered element samarium fell into a place not previously occupied, thus contributing to support the system.

Algebraic Expression of the Periodic Law. — Professor Carnelley has recently studied the periodic law with a view to expressing its numerical relations in the form of an algebraic formula. For reasons which are given in detail in the memoir, an expression of the form

$$A = c(m + \sqrt{v})$$

is adopted, where A represents the atomic weight of the element; c, a constant; m, a member of a series in arithmetical progression, depending upon the horizontal series in the periodic table to which the element belongs; and v, the maximum valence, or the number of the vertical group of which the element is a member.

From a number of approximations Professor Carnelley finds that m is best represented by the value 0 in the lithium-beryllium-boron, etc., horizontal series; by $2\frac{1}{2}$, in the sodium series; 5, in the potassium series; and $8\frac{1}{2}$, 12, $15\frac{1}{2}$, 19, $22\frac{1}{2}$, etc., in the subsequent series. Thus m is a member of an arithmetical series of which the common difference is $2\frac{1}{2}$ for the first three members and $3\frac{1}{2}$ for all the rest. On calculating the values of the constant c from the equation

$$c = \frac{A}{m + \sqrt{v}}$$

for 55 of the elements, the numbers are all found to lie between 6.0 and 7.2, with a mean value of 6.6. In by far the majority of cases the value is much closer to the mean 6.6 than is represented by the two extreme limits; thus in 35 cases the values lie between 6.45 and 6.75. If the number 6.6, therefore, is adopted as the value of c, and the atomic weights of the elements are then calculated from the formula

$$A = 6.6\,(m + \sqrt{v}),$$

the calculated atomic weights thus obtained approximate much more closely to the experimental atomic weights than do the numbers derived from an application of the atomic heat approximation of Dulong and Petit. The number 6.6 at once strikes one as being remarkably near to the celebrated 6.4 of Dulong and Petit, and Professor Carnelley draws the conclusion that there must be a connection between the two. This assumption appears to be supported by several interesting facts.[1]

Prout's Hypothesis. — As early as 1816 the theory was suggested that the atomic weights may be represented by numbers that are exact multiples of that of hydrogen. This led to the further suggestion that possibly hydrogen is a sort of "primordial matter which forms the other elements by successive condensations of itself."

The most critical determinations of the atomic weights seem at present to afford numbers that are not integral

---

[1] Nature, xli. 304.

multiples of that for hydrogen. But in most cases the variations are but slight, and it cannot be declared with certainty that the atomic weights at present held may not be subject to corrections such as will in future afford numbers sustaining Prout's proposition.

Thus recent and careful recalculations of the atomic weights show that "thirty-nine out of sixty-five elements have weights varying each by less than the tenth of a unit from even multiples of the atomic weight of hydrogen." Of the remaining elements, twenty-six have atomic weights that are known to be defectively determined. Thus Prout's hypothesis acquires new interest.

CHAPTER XXIV.

ATOMIC WEIGHT (*continued*).

ELEMENTARY SUBSTANCES AS MOLECULAR.

It is believed that in most cases elementary substances are arranged in groups which may be properly called molecules.

First. This view is sustained by the *volume conditions* of certain chemical unions.

Example a. — Two volumes of hydrogen and two volumes of chlorine combine to form four volumes of hydrochloric acid gas. In accordance with Avogadro's law, two volumes of hydrogen may be considered as having n molecules of hydrogen, and two volumes of chlorine n molecules of chlorine. Then four volumes of hydrochloric acid gas must have $2n$ molecules of the substance represented by HCl. But each of these $2n$ molecules contains at least one atom of hydrogen, a total of $2n$ atoms of hydrogen. But these $2n$ atoms of hydrogen are derived from n molecules of hydrogen. Therefore each molecule of hydrogen has two atoms of hydrogen. By a similar process of reasoning it is apparent that each molecule of chlorine contains two atoms of chlorine.

In certain other cases, elements may be shown to contain four atoms in the molecule.

In some cases, it is true, the number of atoms in a molecule cannot be readily determined.

Example b. — When sulphur burns in oxygen gas to form the molecule SO_2, there is neither increase nor decrease of volume; *i.e.* apparently one molecule of oxygen gas has furnished one molecule of sulphur dioxide, SO_2. This favors the theory that the amount of oxygen gas in the molecule SO_2 is the same as the amount of oxygen gas in one molecule of oxygen. Whence one molecule of oxygen gas appears to consist of two atoms.

When carbon burns in oxygen gas to form the molecule CO_2, there is neither increase nor decrease in volume. By a course of reasoning similar to that just employed with respect to sulphur dioxide, the view then brought forward is sustained by the burning of carbon.

SECOND. This view harmonizes with certain facts of *thermo-chemistry.*

Example a. — Phosphorus burns in the gas nitrogen monoxide, N_2O. It also burns in oxygen gas. Carbon, also, will burn in either of these gases. Now it is observed that more heat is evolved when the combustible burns in nitrogen monoxide than when it burns in oxygen.

According to the molecular theory, oxygen has molecules as well as nitrogen monoxide. The heat afforded by the combustion represents a difference between the true amount of heat evolved by the chemical action and the amount of heat absorbed in decomposing the molecule containing oxygen. The less amount of heat given off by the combustion in pure oxygen suggests that when the phosphorus withdraws an atom of oxygen from its companion atom, more energy is expended than when it draws the oxygen from the companion nitrogen.

Example b. — When the very explosive compound called chloride of nitrogen decomposes, the reaction is as follows: —

$$2\,NCl_3 \; exploded = N_2 + 3\,Cl_2.$$

Now when this operation goes on, 38,100 units of heat are liberated. This extraordinary result is believed to be explainable, on the molecular theory, as due to the fact that the affinity of an atom of nitrogen for another atom of nitrogen, and the affinity of an atom of chlorine for another atom of chlorine, are far greater than is the affinity of the atom of nitrogen for the atoms of chlorine with which it was combined in the explosive substance.

It is true that nitrogen as an elementary gas, and under certain other circumstances, is very inert. But nitrogen is found as a constituent of a very large number of compounds. Some of these are very stable, and the nitrogen holds to the other constituents with great affinity. Ammonia gas (NH_3) is an example. This shows no fundamental lack of affinity in the nitrogen atom. Perhaps the very inertness of the free nitrogen molecule may be due to the affinity binding the two atoms of this molecule together.

THIRD. The facts of *allotropism* accord with this theory.

Some *elementary* substances are capable of undergoing a great change in their properties without any change of weight, and without any addition or withdrawal of other kinds of matter. Thus a given portion of ordinary oxygen may be changed — at least partly — into ozone and back again. So ordinary phosphorus may be changed into red phosphorus and back again. One of the forms of such substances is called *the allotropic form* of it, and this general property of bodies is termed *allotropism*. The view now held is that such modifications represent some change in the number of atoms in the molecule of the element, or some change in their arrangement within the molecule.

Occasionally the term *allotropism* is applied to *compounds* to describe a kind of physical isomerism. Indeed, the three states of aggregation in which a compound may exist may be considered as representing a sort of allotropism.

Many *natural* minerals display a kind of allotropism in their differences from the same substances artificially produced.

The properties of *ozone* support the theory in question. When oxygen changes to ozone, condensation occurs without loss of matter; and when ozone changes to oxygen, corresponding expansion occurs. Ozone is $1\frac{1}{2}$ times as heavy as oxygen; in other words, ozone has the density 24, and oxygen has the density 16. According to Avogadro's law, and the molecular theory of elements, ozone should have the molecular weight 48, and oxygen the molecular weight 32. This means, then, that the molecule of ozone should have three atoms of oxygen, and a molecule of ordinary oxygen should have two atoms of oxygen.

> Now ozone is characterized by two striking properties at first inconsistent; *i.e.* it readily oxidizes certain substances, like metallic silver, adding oxygen to the silver. Again, it readily reduces certain substances like oxide of silver; *i.e.* withdrawing oxygen from the silver. But the inconsistency is immediately explained by the molecular theory. Ozone (O_3), by imparting oxygen to silver, is itself changed to the more stable form O_2; *i.e.* ordinary oxygen. Again, ozone (O_3), in withdrawing oxygen from oxide of silver, at once gives rise to $2\,O_2$; *i.e.* to two molecules of the stable form of oxygen called ordinary oxygen.

FOURTH. This theory affords the best explanation of the properties of certain compounds like *hydrogen dioxide* (H_2O_2). Like ozone, it readily withdraws oxygen from certain substances and it readily adds oxygen to certain substances. The apparent inconsistency is at once explained by the molecular theory. By adding oxygen to certain substances the unstable molecule H_2O_2 is changed to the stable molecule H_2O. On the other hand, in withdrawing oxygen from certain substances, one atom of oxygen withdrawn unites with one atom of oxygen from the H_2O_2, thus producing the stable molecule O_2. But the existence of such a molecule as O_2 is the very point that this discussion is now intended to support.

FIFTH. The superior activity of substances in the

nascent state is best explained by this theory. It is supposed that atoms of a gas when freshly liberated, — in other words, when in the nascent state, — have not yet combined to form molecules of that particular element. They are then more ready to combine with other substances than when, as in their ordinary condition, they have to leave companion atoms to do so.

SIXTH. The theory accords with the facts enunciated in the *law of Charles*.

If the expansion of *compound gases* by increase of heat is due to a forcing apart of a certain number of molecules, the fact that the *elementary gases* obey the same law shows that they must have molecules of the same size as those of compound gases.

If elementary gaseous substances were composed of *single* atoms of smaller size, they should be expected to expand at least twice as much as compound gases by accession of heat.

It must not be supposed that all the substances known to chemists afford satisfactory information as to their molecular structure. This is true of elementary substances, and of compound substances as well. Chemists are still in the dark with respect to the proper molecular formulas of certain elements and compounds. For the present, molecular formulas as well as atomic weights are adopted that are recognized as only approximate, and that are likely to demand revision hereafter.

INDEX.

[THE NUMBERS REFER TO PAGES.]

Acid, acetic, 237, 238.
    formic, 237, 238.
    hydrochloric, 77, 226, 228.
Adhesion between gases and gases, 136.
    between liquids and gases, 134.
    between liquids and liquids, 131.
    between solids and gases, 130.
    between solids and liquids, 114.
    between solids and solids, 114.
Affinity, chemical, 3, 141.
Air, atmospheric, 139.
Allotropism, 251.
Alloys, 128.
    fusible, 49.
Alum, 126.
    crystalline form of, 235.
Amalgams, 129.
Ammonia fountain, 134.
    gas, 227, 230.
    gas, liquefaction of, 59.
    gas, thermolysis of, 146.
Ammonium, 27.
Ampère, A. M., 78, 80, 188, 190.
Analysis, 26.
Andrews, Thos., 20, 35, 58.
Antiseptics, 165.
Arragonite, 236.
Atmosphere, terrestrial, 137.
Atoms, 3, 14, 15, 29, 30, 31, 32.
Attraction, chemical, 168, 170, 187, 188, 189.
Avogadro, 12, 78.

Bacteria, 156, 158, 165, 166.
Balance, 7, 12, 138.
Barium, 3.
Barometer, 8, 9, 10, 69.
Becquerel, 188.
Benzol, 32.
Berthelot, M., 176, 177.
Berthollet, 188.
Berzelius, J. J., 12, 188, 194, 195, 200.
Bismuth crystals, 108.
Botany, 1.
Boyle, 63, 64, 65, 67.
Brodie, 29.
Bromine, 217, 231.
Bunsen, ice calorimeter, 214.

Cadmium, 3.
Cailletet's apparatus, 61.

Calcspar, 236.
Calorimetry, 178, 181.
Calory, 177.
Cane sugar, 124.
Capillarity, 115, 116.
Carbon, 216.
    dioxide, 139.
Carnelley, Thos., 29, 244, 246.
Carré, ice-machine, 51.
Cathetometer, 69.
Charles, J. A. C., 65, 66, 68.
Chemistry, 2.
Chlorine, 216, 231.
Cleavage, 103.
Cohesion, 101, 102.
Colloids, 134.
Compound radicles, 27, 34.
Compounds, specific heat of, 215.
Cooke, J. P., 194.
Corn starch, 153.
Corrosive sublimate, 165.
Critical point, 58, 63, 64.
Crookes, Wm., 30, 38, 39, 41, 194, 244.
Cryohydrates, 128.
Crystals, 109, 110.
Crystallization, 103, 104, 108.

Dalton, John, 6, 29, 196, 200.
Davy, Sir H., 188, 189.
Deleuil, apparatus, 58.
Deliquescence, 122.
Deville, H. St. C., 145.
Dialysis, 132, 133.
Diathermancy, 19.
Diffusiometer, 68.
Dissociation, 49, 50.
Döbereiner's lamp, 131.
Dulong, 213.
Dumas, J. B. A., 12, 188.

Efflorescence, 128.
Eka-aluminium, 246.
Eka-boron, 246.
Electricity, 93, 94, 96, 97.
    related to chemical change, 147.
Electric arc, 96, 147.
    spark, 148.
Electrolysis, 24, 147.
Electroplating, 97, 147.
Elements and compounds, 5.

INDEX.

Elements considered molecular, 249.
    atomic heats of, 215.
    specific heats of, 212.
Energy, 82, 172.
Ether, the, 83.
Ethylene, 237.
Eudiometer, 12.
Eutexia, 49, 123.
Evaporation, 54.
Expansion, 85.

Faraday, M., 29.
Formula, molecular, 225, 234, 236, 237, 238.
Freezing mixtures, 123.

Gas, 23, 36, 57, 67, 209, 212.
Gay-Lussac, 75, 76.
Geissler tubes, 24, 25, 91.
Geology, 1.
Gerhardt, 12.
Germicides, 165.
Gladstone, J. H., 29.
Gmelin, L., 188.
Goniometer, 105.
Graham, Thos., 29, 67, 130.
Granite, 114.
Gravitation, 99.
Gunpowder, 171.
Guthrie, F., 127.

Hannay, 131.
Heat, 42, 82, 83, 144.
    latent, 53, 54, 57.
    specific, 45, 46.
    from chemical combination, 184, 185.
Heterogeneity, 17, 18, 21, 23, 25.
History, natural, 1.
Hoff, J. H. van 't, 118.
Hofmann, A. W., 27, 28.
Hogarth, 131.
Holtz machine, 148.
Hydrargyrum, 3.
Hydrogen, 70.
    dioxide, 252.

Ice calory, 177.
Iceland spar, 21.
Ice-machine, 51, 52, 53, 56.
Infusoria, 157.
Iodine, 183, 217, 231.

Kekulé, 32.
Koch, 155, 156.

Latent heat, 53, 54, 57.
Laurent, 12.
Lavoisier, A. L., 6, 8.
Laws, of definite proportions, 12, 72.
    of Charles, 65.
    of Dulong and Petit, 213.
    of Gay-Lussac, 75.
    of Henry and Dalton, 71.
    of Groshans, 239.
    of insolubility, 171.
    of isomorphism, 104, 234.
    of Mariotte, 63.

Laws, of Mitscherlich, 234.
    of multiple proportions, 12, 73.
    of maximum work, 177.
    of volatility, 172.
Lead, 128.
    boro-silicate of, 23.
Leucomaines, 166.
Light, 20, 22, 86, 87, 88, 146.
Liquid, 36, 52, 59, 60, 132, 134, 239.
Liquefaction, 47, 48, 121.
Lockyer, J. N., 29, 92, 149.

Mallet, J. W., 194.
Mariotte, law, 63.
Marsh gas, 227, 230, 232.
Masses, 3, 6, 99.
Matter, 3, 7, 15, 35, 38, 63, 172.
Matthiesson, 130.
Maxwell, Clerk, 176.
Mechanics, 2.
Mercury, 3.
    as a solvent, 119.
Melting, 43, 46, 47, 48.
Mendeléeff, D. I., 29, 197, 243.
Methyl alcohol, 237.
Meyer, L., 29, 244.
Meyer, V., 146.
Microbes, 150, 155, 157, 159.
Microcrith, 3, 24, 75, 199.
Mills, 29.
Mitscherlich, 104, 234.
Moistening, 115.
Molecule, 3, 4, 15, 28, 78, 101.
Muir, M. M. P., 176, 177.
Mycoderma, aceti, 158, 159.
    vini, 158.

Nascent state, 253.
Naumann, A., 176.
Newlands, J. A. R., 29, 197, 243.
Newton, Sir Isaac, 188.
Nitrogen, oxides of, 73, 74, 232, 233, 236

Occlusion, 130.
Organic compounds, 151, 153.
Organized bodies, 105, 154.
Osmose, 132.
Oxygen, study of, 219.
    unit of atomic weight, 200, 202.
Ozone, 5, 23, 251, 252.

Palladium, 130.
Pasteur, L., 34, 155, 160, 161, 163.
Pattinson, 128.
Periodic law, 243, 246.
Perkin, W. H., 241.
Petit, 213.
Philosophy, natural, 2.
Physics, 2.
Platinum, 130.
Polariscope, 23.
Polarity, 102.
Polarization, 22, 240.
Potassium, 205, 218.
Potassic nitrate, 110.
Potato starch, 152.

Pressure, in liquefaction of gases, 58.
Protagon, 5.
Protyle, 30.
Prout's hypothesis, 247.
Ptomaines, 166.

Quartz, 112.

Radicles, compound, 27, 234.
Radiometer, 40.
Raoult, 38, 118, 241.
Rayleigh, 194.
Refraction, 21, 22.
Regnault, H. V., 211.
Ruhmkorff, 148, 150.

Saccharimeter, 23.
Sodium, 205, 217.
    chloride, 26.
    sulphate, 126.
Salt, crystals of, 111.
Saturation, 122.
Science, 1.
Silver, 127, 136, 218, 223.
Slit, for spectroscope, 89.
Solid, 36, 124.
Solubility, of gases, 71, 134, 135.
    of solids, 118.
Specific heat, 214, 216, 247.
Spectroscope, 86, 87, 89, 90, 91.
Spencer, Herbert, 29.
Spheroidal state, 116.
Starch, 5, 152, 153.
Stas, 194.
Steam, latent heat of, 53.
Sterilizing apparatus, 160, 162.
Stokes, 29.
Sugar, 110, 125.
Sulphur, a study of, 222.
    crystals, 109.
    dioxide, 56, 60, 232.
    density of vapor, 146, 222.
    specific heat of, 223.

Sulphur trioxide, 232.
    use as disinfectant, 164.
Sulphuretted hydrogen, 222.
Synthesis, 26.
Syphon, 71.
Systems, of crystals, 105.

Temperature, 58, 84.
Theory, atomic, 6.
Thermal units, 177.
Thermo-chemistry, 183.
Thermo-dynamics, laws of, 176.
Thermometer, 65, 67, 84, 85.
Thermolysis, 144.
Thomson, Sir Wm., 31.
Thomsen, Julius, 38, 176, 177.
Type compounds, 202, 225.

Uranium, 200.

Vapor, 37.
    density, 211, 213.
    pressure, 52.
Vital processes, 150.

Water, 226, 229.
    as a solvent, 119.
    calory, 177.
    electrolysis of, 148.
    latent heat of, 47.
    spheroidal state of, 117.
    vapor, 77, 219.
Weight, atomic, 193, 196, 199, 204, 209, 245.
    molecular, 231, 239, 241.
Wheat starch, 152.
Wislicenus, 34.
Work, 191.

Yeast plant, 152.

Zero, absolute, 85.
Zoölogy, 1.
Zinc, 3, 120, 168.

www.ingramcontent.com/pod-product-compliance
Lightning Source LLC
Chambersburg PA
CBHW031348230426
43670CB00006B/466